BUY SHEEP
SELL DEER

A Glaswegian
in the Highlands

by

Michael Meighan

**For our dear friends in the North,
Far North and beyond the Far North**

Farewell to the mountains high cover'd with snow,
Farewell to the straths and green valleys below,
Farewell to the forests and wild-hanging woods,
Farewell to the torrents and loud-pouring floods!

Farewell to the Highlands, farewell to the North,
The birth place of Valour, the country of Worth!
Wherever I wander, wherever I rove,
The hills of the Highlands for ever I love.

From Burns: My heart's in the Highlands

First published in 2009 by Booklist Publications

©Michael Meighan 2009

ISBN 978 0 9562 625 0 9

Origination by Booklist Publications
13 Mansewood Court, Edinburgh, EH10 5HA
Printed and bound by the Clydeside Press

Acknowledgements

I would like to thank the following people for their help and assistance. As always to Henry and Janis Mennie, Graham and Lesley MacLennan for the reading of it, for the laughs. To Jill for editing and censoring.

I would like to thank these chums for additional stories:

Frances Nolan
Alan Gartshore
Stephen Meighan
Henry Mennie
Teddy Moseley

Photographs are by Michael Meighan unless indicated otherwise. Thanks to the following for permission to use:

Cruachan Power Station	Scottish Power
Cruachan Dam	Scottish Power
Ferintosh pissoir	Henry Mennie
The Queen going uphill	Henry Mennie
Fort William to Mallaig Steam Train	Neil Macleod, Across Ecosse
Beware of corrugated iron roofs	Henry Mennie
Neil Gunn memorial	Kevin Rae, Creative Commons

Protect my back and get-out clause

This book is based on 25 years in the Highlands and Islands and inevitably things have moved on. I know that many of the hotels and bed and breakfast establishments have vastly improved. In fact they are as good as you would get anywhere in Great Britain now. Well done.

I am sorry if I have innocently offended anyone. I just wanted to record my time in the Highlands and show that there is always a lighter side of life.

As far as I know I have not breached copyright and I apologise if I have inadvertently done so.

Michael Meighan
May 2009

Beware of corrugated iron roofs

Buy Sheep, Sell Deer

Take a look up the rail track

Growing up in Glasgow I can't say that I was that aware of the Highlands and Islands beyond the fact that many people seemed to come to Glasgow to get away from the North. I don't remember people speaking the Gaelic but I do remember the soft lilt of the Western Isles and people talking of home in Barra or wherever.

While I was recently researching a wartime subject I was looking through newspapers of the 40s and 50s. I was struck by the large amount of articles and correspondence then about the Highlands and Islands in central belt newspapers such as The Record, The Glasgow Herald, The Scotsman and the Sunday Post.

If you read any daily or weekly newspaper now in Scotland it would be rare to find anything more than passing reference to things Highlands and Islands. I wondered why and as I did I thought that a book might be a good vehicle for me to record a Glaswegian's 25 years in the Highlands as well as let the good people of the South know what's been going on up there. These are just a few wee stories about the characters and the places and if it encourages you to look up the rail tracks to the North then that will be no bad thing.

When I first went to the Highlands and Islands on the scary old A9 or equally scary A82 it took almost five hours from Edinburgh or Glasgow to Inverness and then another 4 hours to Wick and Thurso. Now the A9 is so good that, barring traffic, the journey to

Inverness is 3 hours and then another 2 to Wick. The trains and inexpensive buses are regular as are the flights. There is even a seaplane. However, many Glaswegians, in my experience, still think that the North is another planet.

I have to make an exception here for that trusty band of senior citizens that has worked out that they can get from Glasgow to Fort William for next to nothing using their Railcard. Well done as the Fort is the doorway to all things wonderful.

So then. What are you waiting for? Take a wee trip with me round the Highlands and Islands and see what you will miss if you don't make the journey.

Now first of all I will explain the title. I heard this from my pal Finlay Matheson (Finlay of the Train) who is a crofter in Lochcarron. I thought that it suited the attitude of the Highlander so well along the lines of 'There are few more impressive sights than a Scotsman on the make' (ref James Barrie in 'What Every Woman Knows').

I will make several references to sheep starting with my first encounter on moving to the Black Isle. We moved into our new house at Alcaig where some people might remember there being a knackery and that was its main claim to fame. Anyway, we had a nice wee cottage and with few neighbours the opportunity to say hello was taken at the earliest opportunity. One Saturday as I worked in the garden I noticed on the road alongside, a flock of sheep milling around. Of course being from the city and being of a tidy disposition and being concerned that the sheep may wander off even further, I sought out the nearest farmer chappy, McDonald I think.

He was of the ruddy faced and aged farmer stock of the Black Isle and came complete with stick and deerstalker hat and walked with me to the road. 'Aye' he said when I pointed out the sheep and from then on he ignored them. 'So you've just moved in?'

I don't know how it happened. In my profession I do have some involvement in and use of deep questioning techniques. But by the end of ten minutes he had fully interrogated me and finally finished with an air of complete satisfaction that he had got to the truth: 'So you are ****** daughter's man?' And so I stayed for years. 'Och yes ****** daughter's man. He's from Glasgow. He used to stay up here for a wee while!' (25 years by the way).

Very recently at a party I met a rather confident but naïve English gent who had moved with his family to the self same area. I asked how long he had lived there. He said 'Two years now and I am not treated as an incomer any more'. He couldn't understand why people were laughing.

I had made minor forays to the Highlands and Islands in my youth but those tended to be more drinking holidays with no real knowledge of where we were staying although I recently unearthed my old youth hostel card with stamps of such exotic places as Glen

Nevis and Carbisdale. There was also the time at the Loch Ard Hotel that is now Altskeith House where we got thrown out for swearing. If three young Glaswegians can't converse in their own language in the Highlands, how do they expect to encourage visitors to return?

Those were the days when we were not allowed into youth hostels if we travelled by car. So much travel was done by hitchhiking or bus. You sometimes stayed where you were dropped. In fact we stayed for a whole weekend in the waiting room at Ardlui Station and I think this is because the lady stationmaster fancied my pal and would invite us in to her cosy room for cups of tea. Sadly while the station is still there the buildings have long gone.

You will see from this work that I have become attached to sheep or certainly used to having them around usually eating our flowers. After 25 years in the Highlands and Islands we moved to Edinburgh and this expectation that sheep would be around I'm afraid followed me. Because each morning in Morningside I would see very posh jeeps with blacked out windows all heading in the one direction just as you would see their muddy cousins in the Highlands. But for the life of me I have not been able to find the Mart. It must be somewhere Behind George Watson's College. There are, however lots of old deers!

We have left the sheep and the deer behind but we still have many friends in the Highlands and Islands and this book is dedicated to them. This is our old neighbour Willie, famous both for his fleeces and for his crooks.

Fleeced at Ferintosh

In Search of The Western Isles!

Towards the West

I spent a great deal of time travelling in and getting around the Highlands and Islands and the very first place that I got to know well was the Western Isles, and particularly Stornoway. Now, just at Stornoway harbour is a building and in that building was a shoe shop. It hardly looked anything from the outside. The windows on either side of a narrow double door were covered in mesh but you could see a few shoes here and there. On passing I thought that I might see if they had a pair for me as I was in need of them.

I squeezed into the shop and squeezed is accurate as going into the shop I could see that there were high shelves on the three internal sides. A counter was between these shelves and the door leaving hardly six square feet of space partly filled by Paw Broon and his old pipe sitting in a battered bentwood chair in the corner of this little area. Talking to Paw Broon and on the other side of the counter was another tweedy looking bent-back old man who eyed me very suspiciously indeed and 'humphed' which I took to mean 'What is it that you want young man? Is it a pair of our finest leathers that you wish to purchase at a very modest price to see you back to the mainland from where you have undoubtedly come?' Aye right.

'Do you have anything in a size 12?' I asked, not so confident now that I was in full exposure to himself and Paw Broon giving me the old quizzical eye and nodding sagely and shaking his head. The old guy looked at me and without a word got hold of the old wooden steps against the high shelves struggling to place them against the shelf where black size 12s might be stored. I thought that I might be late back for lunch as he scaled the ladders one step at a time, taking a two minute break at each step, all the time chatting away in Gaelic to Paw Broon who seemed to be quite happy to frequently take out his pipe from between his gums and aim a big dod of spittle at my feet.

It seemed like an age before the old guy appeared in front of me. I had almost forgotten his mission. From out of a dusty box he duly fetched a pair of 11s in brown. I almost ran from the shop then but courage or stupidity got the better of me. 'Black' I whimpered. Well the look I got would have frozen a penguin. Slowly and with difficulty he got the shoes back into the box and mounted his ladder once again.

Now I wondered whether or not I could endure another pair but he saved me the trouble. Putting the shoes back on the shelf, he came back down again without a box and, completely ignoring me, went back to talking to Paw. I wonder what the word for 'gobsmack' is in Gaelic because I can tell you I was fairly thus. Another bolder Glaswegian may have remonstrated with the pensioner but I weighed in the balance the fact that there were two of them, I was late for the afternoon session of the course and I didn't actually want to end up buying as pair of size 11 brown Tufs. So, my petard being fairly hoisted I went back to the County Hotel to start the second half of the course.

It was there that I said casually to whoever might be listening; 'I nearly bought a pair of shoes down at the harbour'. The silence of a Sunday in Stornoway descended on the group. 'And where would you buy a pair of shoes at the harbour?' Says Kenny. Says I, 'In a wee shop with dirty windows in front of the tea room'. Well there was not one of the 8 who didn't laugh. It transpired that the shop had been owned by three bachelor brothers for many years and as far as anybody could remember, while it had actually been a real shoe shop in distant memory it seemed to now to be little club for elderly male pals. There was only one of the brothers left and it was expected that the club would not last very long.

This was about my first trip to Stornoway and it made me aware that in the Highlands and Islands things are rarely what they seem to be. Because, a number of years later I was relating this story to Donnie Saunders and David Morrison of Isles FM. They told me that in fact this was a very important place for at one time before the imposition of such inconvenient things as Local Authorities, at the back of this shop had been the Parliament, an informal but recognised part of life in Lewis.

This prompted me to think about and write an alternative to what I thought was happening:

One Size fits 'A

So, there was I talking away to Andra about this and that and the ferry and the happenings in Back and the new harbour. Well, you see, in walks this mannie and without a by your own leave at all at all he enquiries in his Kelvinside voice do we not have the likes of shoes in this shop.

Well, the very nerve of the mannie as if Andra and I have all the time in the world to be bothering with him as if he cannot buy the very same shoes in Glasgow. Well it is quite right that Andra looks very suspicious indeed of this object. Who, by the way goes on to ask if we have such a thing as a size 12. Well I can tell you that never in my very own experience has there been such an exotic kind of size of shoe on this island. Perhaps in those barbaric south Islands there may be so.

Anyway, in order to see him off the premises so that Andra and I can get back to the real important things such as the shocking way that the young ladies are dressing in Stornoway these days, I climb the ladder and of course Andra is looking at me as if he was fit to die and me with my arthritics.   Well, of course there is no such a thing as a size 12 and so I comes down again with very nice size 11s and do you know the cheek of the man he asks me if I do not have them at all in black! Well I tell you that that is just about enough and while I took my own good time going up and down, well you know that when I got down he was still there!

Well, Andra and I we looked at each other in wonder and amazement as he must somehow finally have got a hold of himself and disappeared very, very sharply out of the front door. 'Well', said, Andra after a wee whilie. 'Ah doot it must be verry verry hard to get big shoes in Glasgow if he has to come all this way on the boat for them?'

'Indeed' says I.

Do you sell boots?

Yes we do

I'll take two!

At the County

I mention the County Hotel with some tenderness and my pal Chris of Cullicudden on the Black Isle would have backed me up on this. I would often meet Chris, a purveyor of fish, on his island rounds, in the bar of the County Hotel in Stornoway.

When I stayed there it was owned and run by a very nice couple. My first visit assured me of the familiarity with which the visitor might be treated in our Highlands and Islands hotels. On that first occasion I had been berthed in a fairly comfortable double room. I had unpacked my belongings and gone off to have tea. On returning to my room I was waylaid by the lady owner with 'I hope you don't mind but we had a couple arrive and they needed your room so I have moved your bags into a single. It is very nice'. It was at that instant I realised that I should never travel with anything compromising, not that I would do anyway.

But this visit and a wide variety of other incidents persuaded me never to travel without a small toolbox comprising of pliers, screwdriver, sticky tape and some nails. It wasn't a case of 'you might need it', more a case of 'you will definitely need it'. I have had to repair cisterns and replace curtain rods. Do you know the statistics for the number of times that curtains come off the rails when you open them! Doorknobs stick, showers don't work, windows don't open.

On one of my first trips to the County, I thought that I would bathe. Times have moved on apace and you rarely get the enormous deep cast iron baths as they had then. This one was in a bathroom separate to the bedroom and at the top of the stairs. You know how you can see your life flashing in front of your eyes as your end is near? Or so they tell me. Well something like that happened to me in that bathroom. I had let the bath run while I went back to the bedroom to get undressed and fetch my gown and towel. Now the plug on the bath was one of those old-fashioned brass things attached to the bath with a big brass chain.

On returning to the bathroom I felt that there was probably enough hot water in the bath. In fact the water was so hot the room was full of steam. So I went to turn off the big brass tap and it refused to budge. 'Oh bugger' I thought. Out with the toolbox or get the landlord but I had to empty the bath of some hot water first. So I pulled on the chain and it came away but without the plug attached!

So! There I was in an improvised steam room, the water about to come over the top of the bath and cascade over the stairwell into the reception below where I could hear people, for by this time I had the door open and was about to shout for help. But what good would help be to me just then I thought to myself in a moment of lucidity. Back I went into the bathroom, whipped off my towelling gown, wrapped it and my bath towel round my arm and plunged that same arm into the scalding water to grab the butt end of the plug that thankfully came out.

However, I could refer to the old school equation which asks us something about how long it would take to empty a bath if a tap was running at such and such a speed. Well I didn't need to have maths to understand that the bath was filling more quickly than it could empty and I would have to do something about it. I looked around me and there in the corner was a little stool, which I picked up, and I used this to whack the top of the brass tap. The tap had obviously turned but the leg of the poor stool broke and splashed into the water. I was very glad to be able to take my wet towel, wrap it round the tap and turn it off.

No bath! Or at least six inches of hot water in which I lay back absolutely exhausted and waited till the steam abated before checking out the tap which now worked, albeit with a good bit of effort. I put the plug and chain back together and made an exit wrapped in a towel with my wet gown bundled up and still dripping even though I had to wring it out. I confessed to the landlord that I had fallen over his stool in the bathroom but I took fright at saying: 'Not only did I knacker your stool, break your plug, and tap and use up all the hot water in the hotel, not to mention the sopping carpet on the top storey'.

So anyway, that was my bath that I didn't have. And after that harrowing experience I decided to go to the pictures. And, in the Highlands and Islands, going to the pictures can be a strange and wonderful thing, sometimes fostering fortitude under adversity. Anyway that was what it seemed to me to be like in the Seaforth. Now the Seaforth is one of those 60s monstrosity hotels which rarely changed their interiors until well into the 80s. Going to Aviemore in the 1990's used to give you an idea of it.

Anyway, the Seaforth had a large function lounge that doubled as a cinema once or twice a week. Now in the Seaforth this didn't matter so much as the hotel was heated. In Orkney the enormous cinema was only open on a Wednesday and a Saturday, with the heating being put on just before the film. You had to take a hot water bottle, a blanket and a half bottle of Bell's.

Anyway, back to the Seaforth. The advantage of the arrangement was that you could take your pint into the film from the lounge bar. At the bar, while waiting for the film to start about half an hour late, I got chatting to nice chap, away from home like me. The film eventually started and we were ushered in by a young lad who apologised for the delay telling us that the regular projectionist was on holiday. Anyway, we got seated as best we could on the lounge chairs and awaited the start of the film and without any more ado the music came on and the action of the war film came on and then the title:

NOOTALP

14

This was followed by the rest of the opening credits back to front. Of course, as soon as this was noticed by the projectionist, the film came to a shuddering halt and this was followed by the noise of the projectionist obviously trying to manually move the film back. There was a silence until the young man came out to apologise and tell us that if we would like to wait in the bar he would rewind the film and start again. Well we did this and me and my new friend went into the bar to continue our previously interrupted discussion.

After about half an hour we were called back into the film lounge and the film restarted. I was fair enjoying it until the reels had to be changed. I believe that this is normally a fairly automatic process but given the track record it was no surprise that the young chap announced that there would be a short break while he got the second reel ready. By this time my new friend and I were both very mellow and quite ready to be fairly considerate towards the erstwhile projectionist.

So another half hour later, we were called back and sat back to watch the second reel. I was putting it down to the drink as everything did not seem to be where it should be and there was an unsettled rumbling from the crowd. I say crowd in that it started off with about ten and now was down to about four. Well that is a crowd on the islands.

Anyway I thought that I had not been following the plot very well when the film again ground to a halt and, in this instance with a sort of crackle. The young man again came out of his cubicle to apologise and say that he had accidentally put the third reel on second. If anybody wanted to leave now he would give them their money back. Well. It was 11 o'clock. I was scared that they would close up the County so I decided that retreat was the better part of valour. I was eventually to see **Platoon** the whole way through and I thought it a very fine film indeed.

When we first arrived on the Black Isle, the only option was an hour journey to Inverness to the La Scala, this eventually being supplemented by the opening of a small cinema in the Eden Court Theatre, showing mostly non mainstream films.

However, with the building of the Kessock Bridge it is now possible to be in Inverness in 20 minutes to be at the multi-screen cinema complex while the Eden Court continues to show films.

In addition and for more remote areas, there was introduced a novel travelling cinema, the Screen Machine which is built into a large articulated truck and opens up to a largish auditorium and screen. This seems to go down well. According to its introductory explanation, the Screen Machine 'has been especially adapted for the Highlands and Islands to meet the particular requirements of the region's roads and ferry crossings'. This must have been after it's first outing when it got well stuck in the air on a humpback bridge. I don't think 'A Bridge Too Far' was on that night. In its defence it appears to be working well now and travelling round the Highlands and

Islands showing topical films. You can see it on its website if you search for Screen Machine.

They were back to back on the back seat of the bus to Back

And talking of buses, it was only very recently that I was getting the bus from the Retail Park in Inverness into town. I had my large wheelie case with me and was lifting it into the bus when the driver asked:

'Did you pack that case yourself?'

Being of quick wit and repartee, which often gets me into trouble, I was about to say:

'My butler usually does that for me' when I realised from his face that he was either a very good actor or that he was serious. I shut up on the side of caution.

The Uist Tramping Song

Come along, come along, let us foot it out together
Come along, Come along, be it fair or stormy weather
With the hills of home before us, and the purple o' the heather
Let us sing in happy chorus, come along, come along

A Highland Welcome

Snow on Ben Wyvis

'Your welcome was cordial and your cordial is welcome'

(Murdo Matheson of the Highland Fund)

I am glad to say that since I first moved to the Highland and Islands, there has been a dramatic improvement in the standards of both the hotels and the customer care within the tourist industry. There is also a good range of accommodation ranging from excellent campsites and backpackers hostels to excellent mid range hotels like the Cuillin Hills in Portree in Skye or the Alex Hotel in Fort William, which is a great favourite of mine.

In the late 70's, with support from the Highlands and Islands Development Board Tourism Inns Scheme, one company opened hotels in Inverness, Fort William, Ullapool and Wick. These were based on the realisation that most visitors to the Highlands and Islands came by car and there was little modern accommodation to offer them. The Mercury Motor Inns were born and became an institution in the North. However, they hardly went any distance at all in meeting the needs of reasonably priced accommodation and got iced into a time warp that lasted well into the 90s. The same was true of Aviemore, which along with East Kilbride became a sort of suburb of Glasgow. I am glad to say that Aviemore is presently being redeveloped and is as busy as ever. You should go there to go up the ski lift and then onto the Speyside Railway, which starts there. It's brilliant for wee boys of all ages.

And while you are in the area go and see the Ice Factor in Kinlochleven. This is an amazing ice wall that you can climb up or just watch people doing it. Apparently this is 5 times higher than any other in the world. Take your thermals!

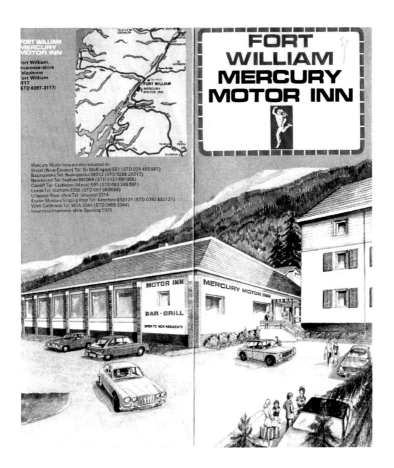

While there is some way to go in the provision of good quality hotels, particularly in Easter Ross, it is not as bad as it once was. At a local Business Association meeting in Dingwall many years ago, a member of the committee suggested that we could hold the next business meeting in a certain small local hotel. When his suggestion was met with a profound silence, and quizzical eyebrows he elaborated: 'It really has improved you know'. Said Hector Munro of Foulis who also attended the meeting; 'It would have to go some'. The next meeting was held in the Council Chambers.

About 20 years ago at a tourist board meeting in Tulloch Castle Hotel in Dingwall, one hotel owner complained about the number of backpackers that were coming to the Highlands and not spending anything. I am glad to say that this shortsighted view has all but disappeared and all visitors are welcome.

This incident and other discoveries showed me that the Highlands and Islands makes very little of its industrial and social heritage. Not that there are no museums or sites of interest. But the little museums are generally run on a shoestring by volunteers and they do a wonderful job. Recently I visited the heritage museum in Broddick on Arran. That little heritage centre is very interesting and is very good value. The Museum of Childhood in Strathpeffer is lovely and so is the Dingwall Museum and the Moray

Motor Museum in Elgin. You should also visit the Wireless Museum if you are in Kirkwall, particularly during the brilliant Orkney Science Festival in September. You can talk to people worldwide as well as seeing a spy's suitcase radio. Brilliant!

What has struck me is the lack of interest in anything of a sizeable nature that will not only attract visitors but which will create employment. In Dingwall for instance, there is the most Northerly canal in Great Britain. It is short certainly but built by Thomas Telford it could easily be a tourist attraction and centre for the display and renovation of old boats as well as an old-fashioned bus station to take tourists round the area. In fact people could come up the canal to shop in the new Tesco in Dingwall using free boats provided by that same company. Why not I say?

On the other hand we have come a very long way in tourism from my first time in the Highlands. The emergence of specialist tourism has been remarkable. From those surfers who brave the waves to divers exploring the wrecks on Orkney. And wildlife has become a major part of this. In Shetland, Seabirds and Seals has carried more than 22,000 visitors from Lerwick around Noss National and Nature Reserve since 1992. This is a fantastic number. The main attraction is the 20,000 strong colony of gannets at the 600 ft high cliffs of Noss. There are also puffins, great skuas, kittiwakes and guillemots.

Taking on water at Aviemore

Islands and pretend islands

In Lerwick Harbour

If you look at a map of Scotland then you will see that the centre of our country is round about Inverness. 'What!' you say 'How can that be if the border is just down the road?' Well the answer is if you take a line from the very south of Galloway to the very tip of Shetland on Unst then you will see the true extent of our fair nation and why the Shetlanders and Orcadians get annoyed about being tucked into the side of the school atlas.

Wonderful windy Shetland is 12 hours away from Aberdeen by ferry. I was fortunate to be able to work there over a number of years. While I don't think that I could be a Shetlander, nevertheless I really took to the Islands. They have many contacts with Scandinavia and they take their Viking heritage very seriously indeed celebrating it with the Up-Helly-Aa Festival.

Now Shetland is unique in its attitude to time. You have probably heard all about the Spanish Maňana syndrome and the fact that they don't have a word for that kind of urgency in the Western Isles.

Well, in Shetland, they take it a bit further. My first experience of this was on my very first trip to Shetland. In Lerwick, I stopped for a sandwich in a little shop. There were a few people in the queue and the rolls were made very fresh and to order. This of course made things very slow and nobody seemed to mind so I just hung on.

To my amazement, a little lad came in and said to one of the two ladies behind the counter: 'Do ye want to buy some raffle tickets for….? Now I would have thought that the response might be to ask the boy to politely return at a different time, preferably when they were closed.

Well they would have none of that. 'Of course' was the response as both the ladies proceeded to write their names on a complete book each.

I was running a training course the next day and during the lunch break I casually mentioned this instance to the participants who were all Shetlanders. Well, with the pleasant and understated contempt that they have for 'soothmoothers' as they called folks from south of Shetland they had a good laugh at my expense and I was treated to other similar stories of latitude in the approach to time.

Apparently if you go into buy a pair of shoes, for example, and you ask for a size 8, they may say that size 8's are 'on the boat' and 'do you want to try a size 9?' I picture this mythical steamer going round and round Shetland filled with the exact sizes of things people actually wanted while the Shetlanders hobbled around with pinched feet and ill-fitting jackets. Just like Stornoway now that I think about it.

In Ross-shire they have an expression: 'first week'. Now I think this variously means the first week that the weather is fine, the first week I can get round to it or whatever you want it to mean it seems to me. This I found out when we decided to get Jake to come and paint our cottage at Alcaig. Now Jake was a lovely chap who was a painter and decorator and who also had a croft, so we should not have expected an immediate response anyway as it was the beginning of the lambing season but he duly arrived kitted out with ladders and stuff.

'Oh' he says 'Michael, if you could put your telephone against the inside of the window then I will know that if it rings twice then I will have to get away back to the lambing'. You have to laugh.

I once travelled to Unst which is the northernmost part of the Shetland Islands. It is an outpost not unlike what you would find in the Yukon I suppose. Two ferries, two drives and there for about two hours before having to get back On the same day. You wonder how Shetlanders ever escape their fate, but they do for the ferry journey to Aberdeen is seen as an integral part of the holiday. So too are the preparations for the journey which also involves asking what relatives and friends need from Marks and Spencer and IKEA. Shopping to the mainland is a necessary part of the whole trip.

I found this out when organising training courses for the Highlands and Islands. Residents on the Islands were more likely to attend events on a Monday and a Friday when they could use their travelling expenses by extending their visits over a weekend. And why not? You certainly needed to allow them a long lunchtime to get to Markies!

I love travelling by ferry and do so when I can in preference to flying. Often I would stay at The Scrabster Hotel after a train journey from Dingwall to Thurso. I would be woken at 5 am by the nice lady and given a full Heelan breakfast before taking the 6 o'clock ferry to Stromness in Orkney. In May it is like sailing in the Greek Islands. The P&O and Calmac ferries are an institution in the Highlands and Islands.

The Calmac ferries in the west are very up to date because in both Gaelic and English they have an announcement for the comfort and safety of passengers just like the airplanes. I think one of these must have been borrowed from Loganair except that they have left a bit in: 'In the event of coming down over water, your life vest will be found under your seat'!

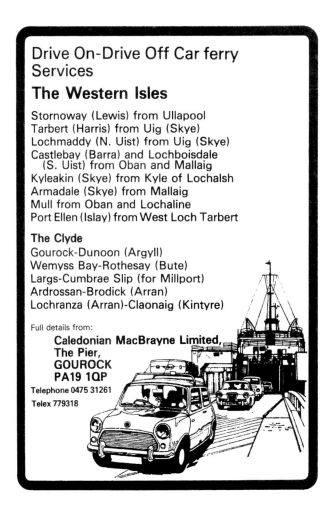

By the way, some years ago I encountered two Glaswegian oilmen in the Lower deck Bar of the Scrabster Hotel. They were complaining about the interfering nature of their Italian boss. One of them said to the other 'Whit's the Latin for spanner anyway?'

I love the ferries in the west and I miss crossing on the old Skye Ferry from Kyle to Kyleakin since they built the new bridge which is quite elegant by the way. At least I did it and have the memories. When the weather is fine there is no better place than Skye and you can still get to Skye by ferry from Mallaig to Armadale all year round and during the summer from Glenelg to Uig. The latter is a traditional turntable ferry run by a community company and the crossing is simply lovely.

Broadford is about midway between Portree and Kyleakin and in Broadford there is a very famous shop called Sutherland's which is now the Co-op. It is the very kind of shop that is central to the communities in the Highlands and Islands and it is where people travelling from Portree to the mainland normally stop for the can of juice and other provisions for the journey.

As a colleague and I did on the way home from a job in Portree. As I wandered round the Co-op I overhead two elderly lady inhabitants of Broadford, complete with headscarves and shopping bags:

''Oh Hello Mary. How are you? ……..Did you hear about Jessie?''.
''I know Mhairi, but she was very lucky. She went fast at the end''.

That's the kind of luck you get in Skye where they suffer tomorrow for today's nice weather.

Oh and while I am on the subject of travelling in Skye I have to remind the prospective visitor to be entirely sure where you are going. I learned this very early on my first trip to Talisker where I was going to visit the famous malt whisky distillery. Now it is a long trip from Dingwall to Skye. The Glaswegian might appreciate this more if I say that it takes as long to get from Inverness to Portree as it does from Inverness to Glasgow. Now. I had located Talisker on my map and headed there but clearly got lost. After fruitless back and forwards in a growing panic I finally saw a mannie in a field who kindly informed me that 'that would be the Talisker on the other side of the island!' Oh dear. You have to be careful of common names. And who would have thought that the Blair Atholl Distillery was not in Blair Atholl, but Pitlochry as I found out one dark rainy night. I mean that would be like saying Queen Street Station was in Hope Street. Not helpful. By the way the Talisker Distillery is a lovely place and well worth a visit.

Oh and did I mention that you should really check the map? You see when I was first sent North to work in the National Hotel in Dingwall I thought it was on an island and I could have made a mistake by going to Tingwall in Shetland, or the one on Orkney. A lucky break then? Of course it was as that is where I met my present wife. (She was my present wife then too, as well, also by the way).

Speed bonnie boat, like a bird on the wing,
Onward, the sailors cry
Carry the lad that's born to be king
Over the sea to Skye

From The Skye Boat Song, Sir Harold Boulton, 1884.

Iona

It's very strange the effect that certain Scottish islands have on people. Just like the thought of Lindisfarne (or Holy Isle). The same is true of Orkney and Iona of which I have heard people talk in awe. The former is a lovely place but you really have to be there in the winter to really appreciate how miserable it can actually be. As far as Iona is concerned, I am sure that is the same but many people make a pilgrimage to Iona on which the famous Iona Community was set up.

I had just bought a cup of tea and a bun in The Camina, the very nice café attached to the Catholic Cathedral of St Mary's in Edinburgh. I had opened an old leaflet on Iona that I had just purchased in a bookshop on Leith Walk. A passing lady saw the leaflet and engaged me in conversation about that holy Isle and how she must really visit it. This illustrates the awe in which it is held. I must go there now and it is about the only Scottish Island that I have not visited.

On the other hand very recently, I overheard three students who were sitting beside me on a train leaving Glasgow Central:

'I'm going to go to Iona next weekend'.
'Whit's Iona?'
'You know – that island where Christianity started'.

Oh Dear!

Oh and by the way, the old leaflet that I had obtained was called 'The Sacred Isle Tour – Oban to Staffa and Iona'. It describes Iona: *'Iona is beautiful, and it is unique. Its atmosphere cannot be equalled anywhere else in the world. It must be seen - **Daily except Sundays'**.* That's the Islands for you

The Black Isle

The Cromarty Rose at Cromarty

There is also a little ferry, the Cromarty Rose, which sails during the summer from Cromarty on the Black Isle to Nigg. This is very convenient for people who live on that side of the Black Isle to get to the Far North. Otherwise it is really only useful for tourists. The fact that it only takes two cars makes getting on quite unpredictable. Undaunted I would always make for the ferry when it was available rather than go the long way round via Invergordon. (By the way, the Black Isle isn't actually an island).

The ferry takes two cars on a mechanical turntable that rotates at the end of the journey to allow you to get off facing front. On the turntable one fine day I shared the available space with a vintage VW camper van of the kind you often see at festivals. The journey over the Cromarty Firth was lovely but uneventful except perhaps for the two dozen or so dolphins and porpoises jumping around us.

At the other side the boatman started the motor for the turntable that whined and turned about three foot before juddering to a halt and refusing to move any further. I could see that the camper van was probably too heavy for the turntable and had caused an imbalance, jamming the mechanism. So there was I stuck on the ferry with very little room to move. I could see that there might be just enough room for me to manoeuvre. The alternative was sailing away to Invergordon in search of a crane.

So very carefully I moved my (brand new) car inch by careful inch by turning the wheels this way and that and I estimated that I had carried out a 53-point turn before being able to move on to the ramp with about an inch on either side. What I hadn't appreciated over the hour that it took to do this was that a huge crowd had gathered at the top of the ramp and I had become the major draw in Cromarty that morning. As I made it on to the ramp and up to the harbour I was greeted by a massive round of applause and smiles.

The Queen going uphill at Cromarty   Henry Mennie

(Please note the Glaswegian on the jet ski)

As we are in Cromarty now I can tell you that the visitor gets quite carried away with the plethora of wildlife in and around the beautiful Black Isle. Serried ranks of photographers can be seen waiting at Chanonry Point near Rosemarkie to photograph the dolphins. And while we are there I can let you know about the poor Brahan Seer. Brahan is an estate just west of Dingwall and a seer is a bit of a prophet. Now they burned this prophet in oil in a barrel for his outrageous prophesies right on Chanonry Point so that he could watch the wildlife while he was burning. Very thoughtful. There is a memorial there to him.

There are now more dolphin watching boats than you can shake a stick at and whenever there is money to be made you will see the competition increasing to such an extent that the dolphins will be chased away and no one will make any money. I can't see any porpoise in it, but you can have a whale of a time.

Talking of which there was a chap in the fishing village of Avoch (Och!) on The Black Isle. He dredged up a live octopus and offered it to the local fishmonger for ten pounds. 'Ah don't know Alex' said the fishmonger whose name was MacDonald 'it disnae look very well'. OK said Alex. 'How about six quid?' By the way, if I have to explain my jokes you should not really be reading this book. Ask for your money back.

Avochie Boaties Graham MacLennan

The Highland Division

Hawker Siddeley HS 748 of the Highland Division

It is very strange that, although I love ferries, I have never ever been on the Ullapool to Stornoway ferry. Many a time I have been in both Ullapool and in Stornoway but always as a destination. Sometime I will do it. All my journeys to the Western Isles have been by air from Inverness or Glasgow and this photo was taken one morning on the 7.25 flight to Stornoway from Inverness

When I first travelled on the 'Highland Division' it was all very casual and you could turn up right before the flight and still get on, not that I would be doing that. All things changed of course when a private security company took over and scanners were introduced.

It was winter and I was off to Stornoway. I had plenty of time so I parked the car and walked over to the main building to check in. Having done so, it occurred to me that I may have left my lights on. We had a little Renault 4 at that time and it didn't have the buzzer that you now get to let you know.

Anyway, I popped outside and saw that my lights were off. Then I went back in and got a coffee to wait for my flight. While I was waiting I looked out of the big picture window and saw a 'plane just beginning to taxi. I wondered where that was off to when it dawned on me that there were no other flights at that time. I raced to the desk where the BA attendant looked at me and said; 'you should be on that flight. Get down to security right away'.

I raced down to the gate but they wouldn't let me on. So I rushed back to the desk and the BA staff member looked at me and said. 'OK just come over here'. So, presumably against all regulations I scrambled over the scales and followed him through a door onto the apron where I could see the 'plane begin to turn. He ran out waving his arms at the pilot. The pilot gave a thumbs-up and the plane gave a sudden jolt as the brakes came on.

The props decelerated while the door opened and the steps came down from the aircraft, the door being behind the wings and props. I was ushered round to the steps and clambered up. The stewardess quickly put me into a seat at the back and closed the door just as the props started up again and the plane taxied round to the runway to take off. I was a bit flustered and a bit embarrassed as I looked up to see everyone on the very full aircraft turn round to see who the 'plane had stopped for. From that day to this I have always thought that 50% of them must have thought that I was a VIP and 50% thought that I was a complete plonker. Changed days now of course. Security is now tighter than it ever was. There are now x-ray scanners where they just gave you an occasional pat down or mostly didn't bother.

The planes have mostly changed now too. In those days, the backbone of The Highland Division was the Hawker Siddeley HS748. I sat there one day with a colleague, Bill who had travelled a bit more than I had at that time. I professed my unease about a nut that seemed to be working its way out of the engine cowling as we were taking off. 'Don't worry' he said 'It works its way back in when we land'. That's comforting. It is true that they were past their best even then but many travellers had more confidence in them than they did in the replacements, the BAE Advanced Turbo Prop (ATP).

We got to know the staff quite well on these Highland Division flights. One day, in passing I mentioned to a stewardess as we landed in Inverness that she would be glad to get finished. 'Not at all' she said, 'I have a long way to go'. It transpired that after the flight from Shetland was complete, the aircraft took off to pick up Services personnel from Manchester, Birmingham, and RAF Brize Norton and to then fly them to Dusseldorf, to be taken on by Lufthansa to Berlin. The HS 748 would then make an overnight return to take back its duties on the Highland Division. She further explained that the 'peace dividend' with the Berlin Wall coming down might have an impact on the number of flights, thus threatening the viability of the Highland Division. In the event this did not happen but it showed me that aircraft are too expensive to be sitting around on the ground for long. It also brought globalisation a bit nearer home.

Henry and the Secretary of State for Scotland

Henry and Douglas of the Dingwall Ramblers (paid up subs £3.00)

Henry, late of Vinicombe Street, off the Byres Road is both a retired senior careers officer and now a sage and a gentleman of leisure. I have known him for many years and I have always been struck by his taste. The fact that he bought our house when we left the Highlands has little to do with it but I do expect him to look after my rare wild orchids.

Now Henry is a man not given to excitement. In fact he is rarely disturbed although his recent protestations about the lack of broadband in the area look to cause concern to his many local friends. Now, it came to pass one day that I happened to be entertaining the Secretary of State for Scotland, the late Donald Dewar, in my offices as part of the campaign to elect Donnie Munro to Parliament on behalf of the Labour Party of Ross, Cromarty and Skye. Ooooooh you don't say?

Henry was not at the meeting but it happened that in popping out for a quick cigarette to Dingwall High Street, I happened to see Henry on the far side of the street striding purposefully. I hailed him and said that I had a visitor that he might like to meet. His response was short and to the point 'OK Michael – I just have to put a line on a sure thing. I'll be with you shortly'. I would point out here that other well-known Labour figures such as Robin Cook shared this affection for the turf.

Henry duly appeared back and joined the throng. I thought that given the good Henry's bona fides, it might not be a bad idea to mix him with the powers that be so I said to Donald ' Would you like to meet the head of the Careers Service for the Highlands?' 'Certainly' he said and I steered Henry in his direction to meet him and talk to him he did for a while before Donald had to depart taking with him 2 detectives, Sam Galbraith as well as MP Lewis Mooney and Donnie Munro.

At that point Henry had to go and see what success he had had with the nags but I will not go into that for fear of embarrassing him further.

Anyway, it was a bit later on, again in the high street when I bumped into the hitherto unfazed Henry. 'You know, he said 'that was unreal. Here's me in my scruffy jeans going to put a line on and the next minute I'm having lunch with the Secretary of State for Scotland. You couldn't buy the experience'. (I charged him there and then).

And by the way I was reminded by my pal Mary McDonald that the week following the visit the chicken and sweet corn soup that she provided during his visit got a mention by him in the House of Commons when he was relating his visit to the North. There are few mentions of soup in Hansard so well done Mary. Donald loved his food by the way. Donald was a fine man. It was a privilege to have known him.

Now, while we are on the subject of Dingwall I must tell you about the visit of the Glasgow Lord Provost, Tommy Dingwall to Dingwall. I got a call from Rankin to say that Tommy had been invited to a reception at the home of Ross County, Victoria Park. Now everyone knows that the Lord Provost of Glasgow travels in the famous Rolls Royce G1.

In due course, Tommy, his wife Grace, chauffeur and two men in black, good friends of Tommy, turned up at Tulloch Castle for the reception. I am sure that the good people of Dingwall and Ross County had no idea that the two bodyguards were hingers-on, good Celtic supporters and freeloaders and their only employment with Glasgow City was in their capacity as teachers. Questions should be asked in the hoose.

Anyway, a good time was had by all and I joined Rankin, Fleming, Tommy and chauffeur for post-celebration drinks in the bar as well as breakfast in the hotel. Following a wee walk round Dingwall to see the canal; it was time for Tommy and entourage to leave. From the front of the National Hotel the Glasgow official car with flags fluttering, set off on its journey back to Glasgow.

At that point, a certain local worthy passed by and stopped as I was waving 'bye' to the High Heid Yin of Glasgow. 'I once met the Lord Provost of Paisley' he said.

Tommy left me with a gift – a scarf in the new Glasgow Tartan. My mother still has it. Nice to have met you Tommy.

Caithness

Wick 1985 M Meighan

There are two well-known hotels in Thurso, which I have stayed in very regularly; the Pentland and the St Clair. Like many hotels in the North, the Pentland has been renovated in recent years. Before that it was very old fashioned with a reception desk closed off with a wooden roller shutter at the bottom of the stairs. At 5.00 in the morning the fire alarm went off. Unexpectedly, my life support systems were in place and I was clearly first down the stairs in my pyjamas and dressing gown. There was no one else around so I was the first and only person at the foot of the stairs to hear the night porter on the Tannoy to all the rooms and in a strong Caithness accent pronouncing: 'It is all right ladies and chentlemen. It is only the toast going on fire. There is no cause for alarm'. So that was fine except that he hadn't considered that the fire alarm going off had cleared the rooms and what I found was a stairway full of people on the way down!

Now the St Clair is a similar but smaller hotel just along the road. There was one incident that caused me some consternation and embarrassment. This was a pool of water that kept appearing in the corner of the bathroom. Yes? Let me finish!

I was due to stay in that hotel for a week and was placed in a fairly typical dismal Highland room where inevitably the light bulbs did not work. Anyway, I believe that I may have had a late night carousing with some travelling salesmen and their ilk in the bar of the Pentland Hotel. It was with some confusion that, in going about my morning ablutions, I spotted the substantial pool of water in the corner.

Wick 1985 M Meighan

Now I may have had a few but I really couldn't remember such an event as might have happened. Anyway, I spent some time scratching my head wondering where the water may have come from and finally gave up, mopping it up with a towel that I hung on the shower over the bath. I said nothing of this at breakfast.

Now then, on returning to my room that night, I first of all checked the floor of the bathroom and was relieved to find it dry. So I passed that night and having to rise early for my offices I was astonished to find another pool of water. A fresh towel had been supplied. This I used to repeat the process of mopping up. I then took a chair from the bedroom and thoroughly examined the ceiling to see if there was an obvious hole but could see nothing.

This was the same scenario for the next two nights. On the fifth night I had to get up a bit earlier and five o'clock found me at the loo just as I could hear the distant sound of the central heating coming on and a great clanking taking place in the pipes. It was to my astonishment and relief (appropriately) that I heard a hissing sound and there, high in the corner from the back of a metal pipe going up the wall I spied a thin jet of water shooting up to bounce on the ceiling and cascade onto the bare linoleum. Mystery solved. It was with a clear conscience that I was able to report said problem to the front desk.

Fisherfolk at Wick Harbour 1985 M Meighan

Now the good lady there was quite busy with other departing souls that morning but took the time to look at me quizzically. 'You have found out how a pool of water has been made in the corner of the bathroom? Thank you very much indeed!'. It was at that moment that I realised that one should leave well enough alone for who knows how long that pool of water had been presenting itself to the confused and unsuspecting traveller who either had a choice of leaving it or mopping it up. I suspect that both may have been done over the years while many guests had come and gone. Why had I been the one to bother?

The Harbourmaster, Wick 1985 M Meighan

There is one thing that I will say however, is that the general condition of hotels in the Highlands and Islands has improved. In no small measure this may be attributable to the myriad numbers of civil servants and other similar itinerants who travel the Highlands and Islands. On the other hand we have seen many a local hotel be taken over by coach and travel companies to service their bus tours from the Midlands of England and further South. The Dalmally Hotel, the Royal in Ullapool, the Crianlarich Hotel. Such hotels have become short-term holiday homes for elderly travellers. There were even stories that certain social services departments in English counties were packing elderly people off on Scottish coach tours because they had no accommodation available in their own shires. It was also with some astonishment and scepticism, I might add, that I was told that all of these coach tours carry cardboard coffins in the event that a traveller might knock on heavens door having once seen the beauty of the Highlands. See Skye and die!

Now Dunnet is a beautiful place, north of Castletown on the North Caithness Coast. I was travelling to The Far North around 1985 and it was suggested to me by a fellow traveller that I could do no better than to sign into The Northern Sands Hotel for a night. This sounded like a good idea at the time and I wondered later why she had given me such a bum steer. What had I done wrong? How had I offended her?

The Hotel had obviously been one of those shooting lodges for Southern industrialists who bought these kinds of hooses without realising how far they were to get to and just used them at the Glasgow Fair. The Highlands and Islands are full of them. Some of them are still around and rotting. Many have disappeared completely. I am glad to say that many of these large houses are now in demand by groups and they can be very nice as witnessed by my very recent stay at Altskeith which is what used to be the Loch Ard Hotel, courtesy of our friends Alan and Wendy Price. In fact there is a Black Isle Company called Large Holiday Houses that rents these muckle piles and castles out to the owners. It's a brilliant idea for Hogmanay.

By the way, as I write this I am listening to Hamish McCunn's overture: 'The Land of the Mountain and the Flood' which goes extremely well with the idea of a great baronial mansion set amid the glens and overlooking a great sea loch. You should listen to it. He is a greatly underrated Scottish Composer whose music could be heard on the original Doctor Finlay's Casebook starring Bill Simpson, Andrew Cruikshank and Barbara Mullen 'Come away in Janet'.

Anyway, I digress. Back to The Northern Sands Hotel which was at that time typical of the family run hotel but run on a shoestring. Normally there would be the threadbare carpets and chairs, the plywood wardrobe, the shared bathroom with the cast iron bath, the public bar with the outside, or very nearly outside toilet with dripping walls. And if I remember they all had cisterns that ran continuously.

I found the gothic pile, now feeling strangely nervous as if I had maybe seen it before in a film. If not Psycho, then the Adams Family. The lady who greeted me invited me to sign in and I was to say the least a bit confused as I had been told by my hotel critic that it was very popular. If so then how come the last person had been there two months before me?

Anyway, nothing daunted, I followed the lady to my room. Oh dear. I must have gone back to the 50's or even earlier. The room was small and narrow, with very faded chintz wallpaper, an enormous cast-iron radiator and cast-iron bed. I thanked the lady very unassertively and slumped back onto the bed whose old springs screeched under my weight.

I turned back the covers and there was no mistaking the damp smell of the sheets that were clean but had obviously been on the bed for some time. I don't normally suffer from asthma but I just imagined my wife exploding and that set me off. I started wheezing at the smell.

I thought that I would check out the bathroom and maybe have a nice hot bath. Although after a few years I had given up the hope of ever doing such a novel thing in a Highland hotel.

I opened the bathroom door and jings, crivens and help ma Roberto, what a surprise. The bathroom was huge and done out in an ultramodern Italian style with fully tiled and mirrored walls, chrome wash basins and all finished in a deep blue and illuminated with sunken spot lamps.

Totally incongruous and amazing except that it was covered in dust and clearly had not been used for some time. It was later explained to me that the hotel had belonged to a local family of Italian descent. Given that every other bit of the hotel was in 50's Highland they must have run out of money after doing up my loo. And no wonder! Except perhaps all the other loos looked like that? I never thought to ask.

You know that the Towrist Board spends a lot of time trying to attract people to the Highlands and Islands using specialist holidays like golf, fishing and there was even an idea for 'pink' holidays. I am afraid that that got a bit of undue attention from the media as you might expect and was blown right out of the water by the B&B owner near Gairloch who refused to have homosexuals over the threshold. Look you can be exclusive or inclusive – which do you want?

I think that the very best thing that they could do is to start tours of eccentric hotels. I can steer them in that direction.

Nice Grocer at Wick Harbour

By the way, before I leave Caithness I would like to tell you about the little known fact that the inventor of the electric clock, Alexander Bain was born near Watten and a replica of his clock can be seen in the local hall in that village which is between Wick and Thurso. And you can see here in this picture of Wick a plate commemorating where the very same man served his apprenticeship as a clockmaker with John Sellar. It is on the wall to the right of the pend. I believe that there is now also a pub in Market Square with his name.

Wick 1985 M Meighan

Alexander Bain went on to patent the basics of the fax machine and installed the first telegraph lines on the railway between Edinburgh and Glasgow. He died in 1877 and was buried in Kirkintilloch, from where sprung Tom Johnston, instigator of the Hydro Dams who was born in 1881.

The Road to the Isles

My chum Frances Nolan reminded me of her exploits in travelling from Aberdeenshire to Uig in Skye by way of the Lecht and Mallaig in January. This would be classed among the great nomadic journeys of our times. They were on their way to do some filming for a training video. Those of you who know the area will also know that this is not a journey to be undertaken lightly, particularly in the snow. However, they had been booked into a hotel in or near Corpach and had been assured that there would be food to meet them when they arrived. The secretary, when phoning said that she had heard the sound of a crackling fire in the background. It was at ten of the clock when they arrived having spent the last difficult hour considering the warmth of a cosy lounge bar and wondering whether it would be the fish and chips or the steak, or the steak pie. Yes of course, you have guessed it. It was cold and miserable and the barman knew nothing about them. But 'I can do you hot pies'. I assume that the crackling noise was that of a crisp packet being opened. I am afraid the scenario used to be fairly typical.

And while I am in that area, Frances helped me remember that it was possibly in that same howff that my brother was persuaded to stop in order to assist a bunch of Lords refresh themselves with large gins and tonic. Sounds pretentious. Possibly, but in his MOD days my brother Stephen really was escorting a bunch of Lords and Members of Parliament to visit military bases in the West Highlands. After a day's wanderings in a cold February they felt that the sun had split the yardarm so had demanded that the bus stop at the nearest hostelry.

Now, I have realised that I should take a moment here to explain that my dear brother, while speaking many languages, has not a bit of the Gaelic. The MOD in question being the real one, The Ministry of Defence, not the Gaelic cultural movement of the same initials.

Anyway, they were well settled into the hotel bar and getting quite used to staying out of the cold. And by the way it is a hostelry, strangely of which I have not partaken, not even a glimpse round the door so I am merely a recounter of others tales. Presumably, Stephen's mobile was out of range in those hilly areas as the phone in the bar rang and was answered by the very wary barman who was also presumably the dispenser of pie.

'Is there a Stephen Meighan in the bar please' was the stout call of the barman. Stephen went up to the bar to take the phone. 'Who is it?' he said to the barman. 'The Palace of Westminster' was the retort said in such a fashion as to receive an 'Aye right! And I'm the Lord Mayor of London'.

It turned out that there had been a three-line whip which sounds painful but is another name for the calling of all Members of Parliament to a very important vote called in the House of Commons that night. The Whips were calling in every MP that they could including those chasing sheep up a Glen in the Highlands to be shufftied pretty quick on their bus with a police escort back to a military flight from Inverness to RAF Northolt.

Anyway, The MPs had to go but the Lords were finely ensconced in the warm bar and were reluctant to move. The colonel in charge of the military was in full flow as it had been a while since he had been able to converse with anyone of his own class. Stephen realised that it was a toss-up between his job if he allowed the party to stay, and the very important Gas Privatisation Bill. Apparently the airplane was on a tight schedule and he only had a very limited time to get the party to Inverness. 'The bus leaves in 4 minutes if you want to be on it' he said. What power!

The MPs had no choice but the Lords were quite happy to stay. But if they didn't get on the bus they were going to find it very difficult get back to London. So reluctantly they gathered themselves together and mounted the bus to the annoyance of the Colonel who was the one used to giving orders. The story was in the Guardian the following day. The gas bill went through and Stephen kept his job. They probably still talk about it in Corpach.

Just a wee interlude here to mention Bernard O'Hagan, Frances's partner, who sadly passed away in 2007. Bernard was the joint proprietor of Redshank kiltmakers in Inverary who mentioned a story about a highland hotel which I did take with a pinch of salt. He and Frances had being staying there and Frances had gone downstairs while Bernard stayed to look after the dog. While he was there a chap knocked on the door to say he had come to service the room. 'That's all right' said Bernard. 'Thanks' said the chap, 'but it's against my religion to do it while there's a dog in the room'. 'Ok' says Bernard. 'I'll just go downstairs'.

And this is where he met Frances who asked him why he had brought the dog down. 'And what religion are we talking about here Bernard?' Asked Frances.

And the big hotels sometimes have a story to tell like the time I was with a party in the cocktail bar of the Dunblane Hydro Hotel. I have delicate teeth so I asked the barperson if the olives on the bar were stoned. She looked at me blankly and asked her colleague. 'Are these olives stoned?' 'Of course they are' she said as I put one into my mouth. Opening one up she said 'there's the stone' just as my tooth was breaking!

While we are talking of secret bases in the West Highlands I must tell you a wee story about BUTEC in Kyle. That is the British Underwater Testing and Evaluation Centre where they carry out testing of missiles and torpedoes.

'Do you know where Kyle of Lochalsh station is?'
'Aye. Just to the left after the secret base!'

Anyway at BUTEC I was supporting a number of managers undertaking a training programme. One of the candidates was in charge of missile recovery operations on a boat after test firings. He was having difficulty understanding the practical assignment that had been set for him. I explained that I thought he was reading too much into it.

'It's not rocket science' I said.

'Actually it is' he said.

And so it was.

Westering Home

Tell me o' lands o' the Orient gay
Speak of the riches and joys o' Cathay
Eh, but it's grand to be walkin' ilk day
To find yourself nearer to Isla.

And it's westering home, and a song in the air,
Light in the eye, and it's goodbye to care.
Laughter o' love and a welcoming there
Isle of my heart, my own one.

Where are the folk like the folk o' the west?
Canty, and couthy, and kindly, the best.
There I would hie me and there I would rest
At hame wi' my ain folk in Isla.

And it's westering home, and a song in the air,
Light in the eye, and it's goodbye to care.
Laughter o' love and a welcoming there
Isle of my heart, my own one.

Anonymous

Happy Campers

Happy campers at The Outsider

Sometimes I just got fed up with hotels and bed and breakfasts and I could be found in various campsites throughout the Highlands and Islands, getting up in the morning, showering and putting on my suit to go and meet a client.

One of my favourite spots is the municipal site in Thurso, overlooking the bay, which by the way, is one of the top surfing beaches in Europe.

So one quiet June found me in a far corner of the campsite putting up my little tent and settling in. Now this wasn't difficult as the campsite wasn't busy at all. So there was I with my little camping stove and billy-can making a cup of tea after a long hard day in Wick.

I heard a very strange noise. No. It was definitely a lorry noise, but one that I had not heard for a very long time. Not possibly since the time of the famous Alness Vintage Vehicle Rally. Hoving into view into the campsite came a bus. If I tell you that it was the likes of bus normally associated with the New Age and with suchlike camp dwellers you may understand. It was a single-decker of various faded hues with windows bestrewn with curtains and other soft furnishings, bags, clothing and whatever else.

It was a bus unlike any that I had seen and I saw right away that it must have been foreign given the strange number plate as well as the overall design. Actually, when I think about it, it probably was a cousin of the bus to Melbost from Stornoway seen in an earlier chapter (with seagull).

Now I was at that stage rather amused at this as anything to occupy one's time in the Far North is very welcome indeed. In fact, I with little interest in birds may have been found at the local birdwatchers AGM and presentations and other such-like not to be missed events all displayed in the what's on columns of the Caithness Courier or Stornoway Gazette. And that reminds me of the Royal Society for the Protection of Birds (RSPB) sticker that I saw on the car windscreen of a grouse shooter in Kinloch Rannoch. Very strange. Anyway, I digress.

The bus roared into my corner of the campsite and stopped not feet from my little bivouac. There with a great bellow and bang it came to a solid and resounding stop. At which point it decanted in a steady stream, what must have been of about 30 bodies all talking in a strange dialect. Being somewhat versed in European languages I thought that I may have been able to discern the origins of the motley crew who had all the appearances of post-war Eastern Europeans.

I imagined that this is what they must have been. So there was I sitting, fairly glaicitly, I might add looking at these people getting out of the bus and I have to say with a little bit of alarm, encroaching upon my personal tenting space. Around the bottom of the bus was a row of doors that swung open outwards. From these cupboards in what seemed to be a well-orchestrated and practiced formula, the party proceeded to decant tents, barbecues, bikes, things, more things and then more things. I use 'things' in the Irish way of using 'yokes' as I had no earthly idea what was being unloaded from the bus. This may have been because of my sudden alarm.

I should have said that although there was I sitting in full view of this bus and its arriving occupants, not one of them said a word to me as they began to set up camp. Neither did they do so when they began to circle me with their tents, getting ever and ever nearer. It wasn't that they were quiet. They jabbered away to one another in their own language. Fairly soon I was surrounded. It was as if I wasn't there. With hindsight I supposed that this must have been what camping in their own country was like.

Anyway, with some embarrassment I upped sticks and re-camped myself in another quiet part of the campsite from where I could hear my old Eastern chums carousing well into the night. This of course, was to prepare me for my first musical festival where anyone camps anywhere. I was privileged to attend my first one at The Outsider Festival in Rothiemurchas. This will be every two years. It is brilliant and you can take the weans.

Willie Logan and the Hydro Dams

Loch Luichart Dam

You may know that I stood for the Westminster Parliament for The Far North. 'The Far North' being the term for the counties of Caithness and Sutherland. While I was preparing for what became a hard fight in which I came second, I became interested in a wide range of issues that had hitherto escaped me. One of these was the debt to which the Highlands and Islands owes to the North of Scotland Hydro-Electric Board. It was the strangest thing to find that, although the dams of the 'Hydro' have been around since the fifties and sixties the contribution that they have made to the Highlands has only recently been recognised, particularly in The Hydro Boys by Emma Wood and The Dam Builders by Jim Miller.

You can't escape the dams and the power stations, the pylons and the lines but they have become as much a part of the Highlands landscape as red deer and sheep.

At this very moment there is an ongoing hot debate about whether or not there should be wind farms throughout the Highlands. Naturally enough there are the 'not in my back yard' types who simply don't want them near but I have recognised over the years that sacrifice by a few has given enormous benefits to many. The building of the dams was the first major economic boost to the Highlands and Islands and has made it what it is today.

The Hydro-electric scheme came out of the need to provide electric power in the forties and fifties and was driven through relentlessly against major resistance of landowners by Tom Johnston, the first Labour Secretary of State for Scotland after the Second World War. His vision was to produce one of the greatest engineering feats of the Twentieth Century as well as make household names of such entrepreneurs as Willie Logan.

During my campaign for parliament that stretched over a year from when I was selected, I was struck by the lack of knowledge that people had about the Labour Party's achievements in the Highlands and Islands. Not only was there little known about Tom Johnston but few people were aware that the Labour Government of 1945 had given them the North of Scotland Hydro-Electric Board, had revitalised the Forestry Commission and had also started the Highlands and Islands Development Board. All of these were to have enormous impact on the development of the Highlands and Islands.

All over Scotland, wherever there is construction work in progress, you will see the name LOGAN. New bridges, motorways, building work of every kind—Logan Construction is playing a big part in shaping the Scotland of tomorrow. So remember, next time you see the name LOGAN, it's a sign that things really ARE happening.

The Building of the Hydro dams was to produce personalities and heroes. Among these was Willie Logan who was instrumental in the story of the dams as well as starting an independent airline company, Loganair.

Long before I moved to the Highlands I worked for Willie Logan on the construction of the main sewer system under the new M8 Motorway through Glasgow. As far as I remember most people that I worked with had a soft spot for Willie who was seen as a fair man and something of a folk hero. They had an expression for him showing their commitment to him: 'Horse on for Logan'.

These included the hard working and hard drinking Irishmen. The Irish navvies were hard men and many's the time I would see them turn up to work in a suit, stepping out of a taxi, get into cement encrusted jeans and go down below to the compressed air tunnel. There they would stay for 8 hours at a time, digging sand and stopping very briefly to eat sandwiches and scoop tea out of a large enamel pail that was one of my jobs to bring to them.

These were the same navvies who built the dams and who accepted the inherent dangers in the business. They talked about the 'bone rot' or asked 'How much did so and so get for that arm/leg there?'

Willie and his wife lived in Dingwall in a relatively modest house on Kinnairdie Brae. My wife, also from Dingwall, tells me that every year he and his wife would give a Halloween party complete with dooking for apples. She told me that he was a very pleasant man. Unfortunately he was killed tragically in the crash of a small aeroplane on his return to Inverness from a meeting in England. Ironically it was not a Loganair plane as none were available.

Loganair has played a major part in the development of the dams as well as of the Highlands and Islands.

Loganair is now a franchise partner of Flybe although there are Loganair signs on the planes. I have a soft spot for the planes as they have given me many an interesting experience.

Britten Norman Islander at Orkney Airport

Before we leave the Hydro-Electric schemes you must go and see the amazing Cruachan Power Station near Oban. This powerhouse is buried a kilometre below the ground and inside is a cavern high enough to house Tower Bridge in London. In here the giant turbines convert water into electricity. You will also see the story of the dams. It is brilliant. You can get the Oban train from Glasgow and get off at Falls of Cruachan station.

The Cruachan Dam

I have lost my luggage a couple of times but the strangest event was on a trip to Kirkwall via Wick from Inverness in a Loganair Britten-Norman Islander. Now the Islander was for many years the backbone of the Loganair fleet. It only held about 6 people but it took off and landed in very short order and was really very exciting and I flew in it often.

The route to Wick from Inverness if you go in a straight line goes right across the RAF bombing range at Tain. If the range is not in use then you can fly directly across the range rather than do a dog-leg. Permission to do this has to be sought from RAF Kinloss control. You can save some time doing it this way.

Now many people in the Eastern Highlands know about the bombing range as a whole range of bombers and jets travel from air bases throughout Great Britain and beyond to practice bombing. Unfortunately they sometimes lose their bombs in places other than the range. The biggest problem is the noise caused by low-flying jets who are supposed to obey orders to fly above a certain height and often do not. Besides scaring the shit out of you it frightens horses and makes sheep abort. Someone should do something.

Anyway. I happened to be the only passenger. While this rarely happened it was quite usual to fly with only one or two others. The pilot invited me up besides him and offered me the spare boxes (earphones) to listen in as we flew. Flying can be a very lonely business if you don't have a co-pilot.

We took off from Inverness and having cleared air traffic control the pilot tuned into RAF Kinloss and was given permission to head over the bombing range as it was all clear. It was a beautiful day and we chatted as we crossed the Black Isle and headed northwards with the Dornoch Firth ahead.

''Kinloss Control to Loganair G-BFX - Standby'.

''Kinloss Control to Loganair G-BFX. Please disregard previous. Inbound fast aircraft on heading ….. should be in your sight in ……Please acknowledge. Repeat……'.

'Affirmative Kinloss Control'.

'Did you understand that?' the pilot said to me with just a hint of concern in his voice.

'No' I said. What does it mean?

'There is a military jet heading in this direction at the same height as us. Just keep your eyes peeled can you?'

There appeared to be not a lot we could do. Someone had made a mess of communications and we seemed to be at the blunt end of it. Anyway the journey to Wick passed uneventfully and during that quiet period the pilot told me that communications between military and civil air traffic control wasn't just as good as it should have been. A bit of an understatement I would have thought.

It was with a bit of a relief that we landed in Kirkwall and I said goodbye to the pilot. I may have forgotten about the incident completely except that I was staying at the Foveran Hotel near Kirkwall and while I was tucking into a steak with a glass of red wine there was an explosion out at sea. The next morning we were to discover that a jet on military exercises had crashed with the loss of the pilot and navigator.

It has always been a mystery to me why more use has not been made of air services in the Highlands and Islands given the remoteness and inaccessibility of communities and businesses. While there is one ambulance Eurocopter based in Inverness and others provide air-sea rescue services, there is ample scope for the better use of airplanes particularly for such things as patient transfer. It has been shown that this is also far safer for the patient if he-she has to be taken to hospital overnight in the winter. And let's get the myth of the effects on the environment out of the way. Executive jets are obviously costly and polluting but if 10 people drive from Inverness to Wick it currently takes 2 hours on a good day. A Shorts 360 aircraft with a full load flying from Inverness takes about thirty minutes and saves on road use.

I always though that if Willie Logan had survived then we would have seen more developments in air travel, not only throughout the Highlands and Islands but also connecting with elsewhere in Great Britain. I am particularly glad to see that there are presently moves to re-open both Oban and Broadford airstrips to commercial traffic.

In the meantime, a brilliant development is the new floatplane service from the centre of Glasgow to Oban that takes 1 hour rather then the four hours by train. Loch Lomond Seaplanes launched the service in 2007 using a 9-seater Cessna. Captain David West is the company's founder and he has ambitious plans to open up further routes using 14-19 seater aircraft. Very good luck to him particularly if it cuts out some of the vehicle traffic on the dire A85.

Inside the Cruachan Power Station

In the Good Old Zimmer Time

Talking of standing for Parliament which I occasionally may say something about, I have to tell you about Susannah Stone, her good works and zimmers. It so happened that I was asked to attend a hustings with the other candidates at Tain Royal Academy which was packed on the night. If you have seen Richard Powell in the 39 Steps you will have a very good idea of what a hustings is and what a trial it can be. It is intended to give ordinary people a chance to interrogate the prospective candidates. In reality the people who generally turn up are supporters of the various candidates who invariably make life very difficult for the others. If you are in a particularly hostile place for Socialists, say like Lairg then you may get a very rough time.

Anyway. There were about 200 people in the hall and during the questions directed at me a lady asked me why couldn't her mother get a zimmer frame in Caithness General Hospital? Now hustings are different from law courts in that lawyers should never ask a question to which they don't themselves know the answer. At a hustings the idea is that irrespective of anyone in the audience knowing the answer you have to be very sure that the poor candidate that you are questioning doesn't know it at all.

I had just visited the fine Caithness General Hospital and talked to its fine staff. I commended them on the work going on towards a fine new accident and emergency department.

It didn't really occur to me to ask: 'Now that's all very well now medical staff but can you tell me if you have enough Zimmer frames in stock? And what about crutches and bedpans?

So when you are asked the question, the experienced politician waves it off by saying that they will ensure blah...blah...blah. To the Labour Party inexperienced politician talking to a mainly liberal and SNP audience it doesn't really matter what you say. You're going to get bedpanned anyway as I did and I really don't remember what I actually said.

Now at these hustings you normally have to hang around because someone wants to chin you about student grants, bin collections or the free bus pass. On this occasion the first person that came up to me was Susannah Stone. Besides being associated with the cheese company that makes the famous Caboc cheese, she is also mother of the sitting Member of the Scottish Parliament, Jamie Stone a liberal and obviously very closely associated with my adversary, John Thurso also of the Liberal Democratic Party.

What Susannah said to me confirmed that the LibDems were wielding a campaign of love so that I wouldn't fight too fiercely against them as the SNP were doing, hopelessly anyway. I was actually doing quite well and at one stage the Herald had me down to win. I was led to believe that in higher places in both the Labour and Liberal Democrat Parties there was heightened anxiety and a certain level of miff as there had been an agreement that as they had a nice wee relationship in the Scottish Parliament they wouldn't interfere with one another in parliamentary seats in the Highlands and Islands. I can't believe that of politicians, can you?

Meighan makes cabinet

Anyway Susannah started with 'Jamie popped in for his tea tonight and said that you were a very nice person'. (see). Do you know Michael, I think that's it's a great pity that candidates can't really talk from the heart at these hustings any more'.

'I couldn't agree with you more' I said 'what I really wanted to say about zimmers was "Vote for the Labour Party and we will guarantee zimmers at the Caithness General Hospital because the zimmer time is coming!" I wish I had now as it might have given me the publicity I needed in the final days of the campaign. She thought it was funny anyway.

Nice to have met you Susannah.

And talking of hospitals as I was, I must recount my story of my gall bladder as well as dedicate it to all those who suffer from the affliction, including Mary at the Mart. Where would we be without being able to discuss septic tanks and medical conditions in the Highlands and Islands?

Anyway while it is a long story I will only tell you the bit from where I was admitted to Raigmore hospital for a planned gall bladderectomy. It was a Sunday evening and I had been admitted then to prepare for an operation on the following morning. I had never had an operation under anaesthetic before and I felt both quite miserable and apprehensive and certainly not in the mood to converse with anyone.

I was delivered to the small ward where there were four beds, none of which were occupied. There was, however, an elderly gentleman sitting in a wheelchair and it appeared that he had only one leg. I was settled in by my nurse and things explained to me. I then had to make myself as comfortable and pass the time as best I could. I was busying myself with my own affairs, not really wanting to make eye contact with the old chap. I knew I was probably being quite selfish but I really didn't want to have to make conversation.

However, I have to tell you that my innate pleasantness got the better of me and I looked over at him so that I could give him a smile and not appear churlish. He returned my smile and then asked me what I was in for which I told him very shortly. I then made a very big mistake by thinking that I would be polite and return the question expecting a similarly short response.

However, I was greatly mistaken. He proceeded to tell me his tale of woe. Apparently he had arrived at the hospital in Nairn for a minor operation, had contracted MRSA, and had been moved to Edinburgh Royal Infirmary where he had his leg amputated before being moved back to Raigmore. Just at that very moment he was waiting for transport back to Nairn to convalesce. In all he had been out of circulation for about 4 months. I am sure that he meant well and a trouble shared is a trouble halved and all that rubbish. However, I wasn't particularly keen to discuss my problems with him at that moment.

The ambulance people arrived just then and greeted him as an old friend. I thought wryly that I may yet get to meet a lot of new friends if I had to stay in hospital. So he was taken off and I raised the glimmer of a smile to wish him well. Then it was all quiet for about twenty minutes while I digested the story and its possibilities. Then there was a shift change and a male nurse bustled into the room making his presence felt by poking a few magazines and generally tidying up.

He spotted a slipper on the window ledge and this prompted him to look round in a ruminative fashion. He looked at me and said 'hi. Did you see the old gentleman in a wheelchair that was here?' I replied in the positive saying: 'You mean the gentlemen with one leg? Yes – he has just left'.

'You didn't see which leg he had lost?' 'Yes' I said, 'it was the left'. He then looked at the slipper, thought, and then said: 'Oh well he won't be needing this then', throwing the left slipper into the bin! I don't think I slept that night.

And there was also a residual worry. I wondered about the old chap with one leg and why was he allowing them to take him back to the hospital in which he started out fully limbed? Maybe he had been on a positive thinking course. I have these kinds of worries. I should get over it. I felt like the guy in Raigmore hospital who had the stomach problems and had been getting purged daily and was getting awfully tired of it. He finally locked himself in the ward toilet and was refusing to come out. The Nurse Manager finally arrived and pounded on the door. 'Mr Smith, will you come out please?'

'Who's there, friend or enema?'

My thanks to John Curran who told me this while we were working in the Dalmally Hotel.

And if you fancy going to Nairn to convalesce or simply to visit, it has a vintage vehicle rally as well as the famous Nairn International Jazz Festival in August. Nearby Forres has the most brilliant floral displays as well as an interesting wee museum. In between is the lovely Brodie Castle. This is very accessible to all and extremely interesting. There are also lovely walks. Children love it.

Oh and by the way, did you know that Nairn is the fastest town in Scotland?

Naaaaaaaaaaaaairnnnnnnnnn………………

The Dingwall Mart

Now the Highlands can be very traditional and resistant to change. I can explain this by reference to Teddy's Goulash. 'Teddy's at The Mart' is run by our very good friend Teddy Moseley who, along with a very cheery crew, feeds the farming stock of Ross-Shire in the Auction Mart in Dingwall. Her bacon rolls are famous. Plain and wholesome fare but it wasn't always meant to be plain. No one would accuse Teddy of not wanting to serve the Heelan farmer the best of European and World provender. With this in mind, not long after opening she offered up on her blackboard 'Goulash'.

It may have been the finest of local produce, beef from Fraser's or Munro's in Dingwall, vegetables from local farms, but by noon there were no takers, not even a taste. Her assistant Mary suggested that she change the menu board. She did this with trepidation and a nagging doubt; she rubbed out 'Goulash' and replaced it with 'Stew'. It was gone by one o'clock. I don't think that it was national pride or xenophobia, just that the Heelander doesn't quite like to embrace change so quickly. 'It's all very well having a brand new Mart, but let's not get too carried away. It might be slightly European funded but it doesn't mean that our meals all have to be European'.

Although she now tells me that there is a bit of a change in that she now sells decaf coffee as well as 'hippy tea'. One farmer looked at the selection of tea bags at the till as he was paying for his stew at one of the Friday sales. 'Now whit are these?' he said, in a puzzled fashion to Teddy.

Teddy explained that they were peppermint tea and such like. 'Oh well', he said. 'That's quite all right then. I thought that it is these things you get in the chemists and I might be getting asked if I wanted something for the weekend'.

And talking of the Mart, that's one of the biggest changes in the past thirty years. At one time, most largish county towns like Dingwall had their local auction marts with sales being conducted once a week or less. There has been a major rationalisation of the mart system. Gone are most of the ramshackle marts with their old wooden stockades in the middle of town. Gone are the sheep and other droppings in the High Street. Gone are the bulls getting loose and charging into china shops. Except to say that I have just heard about one getting loose and plunging into a private outdoor swimming pool in St Andrews. The stranger thing is why anyone would want an outdoor swimming pool on the East Coast.

The new marts serve larger areas, are mostly in new out-of-town sites and are run using hygienic and technological methods. I saw the brochure for the new Dingwall Mart and they have 200 sales, most of them in Dingwall, but some in satellite sites in places like Lochmaddy, Fort William and Portree.

Of course, many of the changes resulting in closer control of rearing, movement and processing of beasts have been because of the BSE and foot and mouth problems. These issues have probably speeded up rationalisation and developments and this may not be a bad thing as it has pushed against the conservatism of the Heelan farmer. But agriculture in the Highlands and Islands has responded well to new markets and opportunities. All credit to Kenny Mackenzie and his team at Dingwall and United Auctions for embracing the changes required.

Now, for years I have loved the furniture sale, those auctions where you still can bag a great bargain. I have been to salerooms all over Scotland and there is no better than Fraser's Auction Rooms in Dingwall, part of the Dingwall Mart. It is a tradition, a great draw and often a great laugh to hear the gravely voice of Kenny Mackenzie preside over the sale of a pile of curtains, a box of books or a wardrobe.

And they take their professionalism to heart. They are unflinching in the face of adversity and able to take a joke. One Friday afternoon in the busy saleroom, two of the porters lifted up a large television for the room to see. Now it was clear to the audience that the innards of the TV had been removed and replaced with rather a nice montage of 2 stuffed pheasants in their natural habitat. So Ian Tolmie began the bidding 'who'll start me on four pounds for this black and white set?'

Of course he was mystified when the room erupted with laughter. The porters shuffled round with their load so that he could see it. 'I beg your pardon' he said 'who'll start me with £10 on the colour set?'

Recently, Fraser's Auction Rooms has moved to be at the new mart but the sales are still popular and I am sure it will be a tradition that will survive. My own feelings are that the traditional 'furniture' auction should be encouraged as a tourist attraction in its own right.

But what has been a great development is the opening of the Highland Drover Project by the Highland Livestock Heritage Society and based at the Dingwall Mart. This is an important archive and permanent exhibition of literature and artefacts recording the achievements of the Highland cattle breeders and drovers who walked cattle from the North to the South of Britain. Well done to all at the Dingwall Mart.

By the way if you are near Dingwall go and walk in the Brahan Estate where you will see this lovely pool and a very unusual dogs graveyard. You can also go horseriding there.

Evening pond at Brahan, Maryburgh

The fish supper

Talking of food puts me in the mood for a real fish supper like you would get in most good fish and chip shops throughout Glasgow and its environs. Not your namby pamby soggy battered ones that they have sold me in Stornoway and Kyle. Though I have to say The Conon Chippie is not bad and for sheer excellence you have to go the Fishermen's Mission in Lochinver.

Anyway the reason I raise the subject is because I was reminded of an incident in which a fish supper was central to the event. Now this could very well have been me so I am loath to criticise. There was this poor chap whom, in my minds eye, I immediately presumed to be from Glasgow on the basis of 'It takes wan to know wan!' Anyway, that's what I thought when I heard the story.

This was set in my parent-in-law's house in a leafy part of Dingwall. While not being a particular nuisance, one of the other villas had taken to providing a bed and breakfast service for the travelling classes among whom, statistically speaking, would be those whom, having spent some time in the strange assortment of Dingwall pubs to taste local culture be desirous of a nice fish supper before retiring.

This truth found one poor soul who 'worse for wear' weaved and wobbled his way from Donati's fish and chip shop covering the High Street twice by dint of diagonal wanderings.

Anyway, he arrived eventually at what he took to be his rooming house and proceeded through the unlocked door, climbed the stair and turned into what he thought was his room whereupon he was spotted by my mother-in-law who was sitting up in bed reading a book. It is not known who got the bigger fright. The chappies arms and legs went wide, his eyes popped with horror and he took himself down the stairs to disappear with a crash through the door, leaving in his wake, a trail of chips and vinegar-laden paper.

Now it must be explained that in the Highlands and Islands, doors might be left unlocked as a matter of course except at the Glasgow Fair Fortnight of course. On this occasion the door had indeed been left unlocked to allow a return to the fold of my wife's younger brother who was still abroad.

Anyway, silence had again descended after a small clean-up operation and the parents in law returned from whence they had been cruelly ripped and the lights were extinguished.

And I often wondered about this. As the lad was not yet home, the door was still unlocked!

So it was with a bit of relief that my brother-in-law was heard to return, the door quietly opening and him tiptoeing up the stairs. Yes. You guessed it – our bold reveller had returned and a similar scene ensued except that this time, my father in law, always the gentleman followed the chap who was by this time in a heap having rolled down the stair banging into the wall on the way. He led him towards the front door and pointed out to him his true hostelry. At the same time, the son, my brother in law returned looking strangely at this chap who head down, and a bit more sober, headed towards his B & B.

So again quietness descended on Dingwall and all that was left was a lingering smell of fish and a good story. Oh yes, and the large hole in the decorative dado in the wall (look up dado – I had to. It's quite innocent). These are not easily repaired and neither, I assume was the poor chap's knee or other part of his anatomy that made the said indentation.

As I said, it could easily have been me and I could have ended up in a highland jail. This one is in Inverary and worth taking the children for a visit or maybe to leave them.

Sad Glaswegian in jile

A policeman's lot

There was a time before technology when everybody knew the local bobby. Those were the times when just about every village had its own policeman who lived in the Police Station.

In Dingwall, the local bobby was very well known, MacDonald, I think. Before I go on I should tell you that Dingwall has its local dialect and in this dialect, 'gadgie' is the name for a person, often of an eccentric demeanour.

It happened that one day PC MacDonald, in calling into the police station, was despatched to investigate an incident in the new swimming pool. Apparently there had been a call that someone's pet budgerigar had escaped, got into the pool building and was now flying about the roof space.

Constable MacDonald duly arrived at the pool to enquire 'Where's this gadgie that's going wild then?'

This reminds me to tell you that in Dingwall and environs, if you ask someone how they are doing as in 'What's the craic?' you are as like to get one of two responses: 'Oh surviving', or 'You're seeing it'. These of course do not make a great impact on the page but when heard in a Dingwall accent they are both very funny. The Great Andy Stewart made a record called 'The Rumour' in which he catches the Dingwall and other Scottish accents perfectly. You should really listen to it, or come to Dingwall.

One weekend I had had a particularly arduous journey to and from Invergordon in a clapped out hired van from a hire company in the Industrial Estate. I had to leave my car overnight and I was concerned about its safety. The company said that I could leave it in their premises and they would put it outside the next day from where I would collect it.

The gears on the van were knackered and I had to come from Dundee to Inverness via Aberdeen at the rush hour because there was no way that I was going to get the van up the Fettercairn to Banchory or Cockbridge to Tomintoul roads. I got back to Invergordon at midnight, three hours later than I had intended and totally knackered. Alas and alack, the car was still in the garage and I could see it through the window. I had to get my car and not having any contact numbers, I wandered down to the local police station in Invergordon High Street to see if the police could contact the key holders of the hire company.

The police station is in an old house with a garden and I wandered through to the door where I knocked. There was no answer so I turned the handle and wandered into what was a reception area. There was still no one around so I looked around before finally leaving and going over to the phone box across the road.

'Alness Police Station'.

'Hi. I've just called into the Invergordon Police Station and there is no one there'.

'That's right. It's shut'.

'No it isn't'.

'What do you mean?'

'I mean the door is unlocked'.

'Just stay there please and we will send someone along right away'.

And so they did with some haste. Two constables established the fact that the nick had not been broken into and the front door had simply been left unlocked when the staff had vacated through the back door. They also established that nothing had been nicked from the nick and very kindly sorted out my vehicular problems for me rousing a very apologetic woman from her slumbers to open the garage door for me.

While we are in Easter Ross I must tell you of a story told to me by Alistair MacEwan. Now Alistair was one of those original poor souls who dedicated years of their life to implementing the brand new Youth Training Scheme. In many cases YTS was a brilliant start to working life for many young people. Part of the YTS scheme was to allow each young person 13 weeks off the job training. This could consist of many vocational or life courses and we experimented with residential courses that were very successful.

Alistair was talking to a local coal merchant in Easter Ross whose name I am sure was MacDonald. Mr MacDonald was enquiring into the possibility of obtaining the services of a young person for the purposes of humping coal bags. When it was explained to him the requirements of 13 weeks off-the-job training, Mr MacDonald retorted. 'But dae ye no jist have an Alness lout for me?' That's what we were up against.

Now by the way, I do not wish you to think that I have any particular downer on Alness. It is a very pretty town and it has come along way in recent years due to the wonderful community spirit that has helped it achieve 'Scotland in Bloom' and 'Britain in Bloom' several times. Alness was virtually a new town built round the old village to house aluminium smelter and oil workers in the 1960s and 1970's. As the Invergordon Smelter closed and the demand for rigs diminished, many workers moved on leaving a rump of unemployed with the attached consequences.

It has sometimes been difficult for incomers and their children to be assimilated into rural communities and sometimes they have ended up as long-term unemployed. One example was an awfully nice chap, originally from Castlemilk who then lived in Alness. He came on a course I was running for jobseekers on a local authority programme. He turned out to be very participative and keen. The following year he was still on the dole and was on a second identical course that I was running for the Jobcentre. Again he was a very productive and enthusiastic participant, even assisting me control some of the more disruptive elements. He came up to me at the end and said:

'Michael. I fair enjoyed that again. I'll look forward to seeing you the next time'!

You win some you lose some.

Rigs berthed in the Cromarty Firth

Fish

On the Cromarty Firth at Alness

During my election campaign I was able to get Brian Wilson MP and Alistair Morrison MSP to come to the Far North to press the flesh and support me in my campaign. It was arranged that they would visit the local fish processing company Aquascot. This is within the normal process of campaigning.

Now Brian had seen round the factory and we were taken to meet some staff representatives where we engaged in some light political discussion. There was some chat about the rights and wrongs of culling seals to preserve the salmon that after all is the bread and butter of Aquascot. So Brian thought he was on solid ground when he replied to a question that he didn't see what was wrong with doing away with a few seals for the better benefit of the salmon industry. However there was one cove that took extreme exception to Brian's view of seals and proceeded to give him the old verbal thumbs down by way of a bit of a tirade. Brian was slightly surprised and taken aback by the proceedings as he thought he was in safe waters! His fate was sealed.

That being over we all retired to my car where Brian gave an interview to BBC Radio Four about the current problems in an Asian country where several British subjects were due to be tied up and lashed for some offence or other. He being at that time a Minister of State at the Foreign Office, he was duty bound to tell those abroad about

Her Majesty's Government's dissatisfaction if not extreme concern about the proposed lashings of poor British Citizens who had been found guilty of some local crime.

The interview being over we set off and I was reversing the car. Alistair Morrison was next to me and he looked back and said in his Western Isles accent: 'well well, he said and they tell me that a week is a long time in politics but with you it is chust ten minutes'.

What do you mean? Says Brian?

'Well right now you are against a wee bit of lashing of a couple of young lads but ten minutes ago you were all in favour of bashing in the heads of a few poor wee sealies! Shame'.

And talking of Aquascot, which is based in Alness, in Easter Ross, takes me on very neatly to Scottish Farmed Salmon that is one of the great success stories of the Highlands and Islands.

When I first came to the North, fish farming was in its infancy and had a lot of lessons to learn. The people involved had to overcome a wide range of problems including political, environmental, technological and economic. You can read all about these in Eric Macleod's book Kerracher Man.

None of this would have been possible without the creativity, resourcefulness and perseverance of a small group of entrepreneurs such as Denis Overton at Aquascot, Ian Anderson at Strathaird, Gilpin Bradley at Wester Ross and John Kerrison and the late Chris Rae of Corrie Mhor Salmon. As an example of the breed, Chris Rae was the epitome of the fish pioneer. With an ecology degree from Edinburgh University in the 70s he was first employed on one of the first salmon farms in the Highlands and Islands, at Eskadale near Beauly in Inverness-Shire. Chris then left to set up a salmon farm in Somerset, the Great Western Salmon Company that he ran successfully. He returned to the Highlands in the 80s to be at the forefront of seafood export, establishing Osprey Seafoods. He then started his own salmon farm near Plockton in Wester Ross but when his brother-in-law John Kerrison returned from similar ventures abroad, they both set up the successful Corrie Mhor Salmon Company in Kishorn, growing over 15 years to be one of the largest independent smolt (young salmon) producers in the country.

With multinationals now dominating production and marketing, John and Chris received an offer and sold out to Panfish in 2005. Sadly Chris was killed in a tragic road accident in 2007. I knew Chris over the years and met occasionally in his Wester Ross years, on the British Airways flight to Stornoway. On one occasion he sat quite happily with a very large black water tank for his fish farm lodged between the seats in the row in front.

On one occasion we again met on the same flight and on arrival, he simply went round to the back of the plane to where the luggage lorry was. He picked out his bag and with me following in some awe just said 'No point in hanging about like' and we left to be the first in the queue for the only taxi.

This risk-taking attitude typified his breed and to me he was one of the true heroes of the Highlands and Islands who pushed to establish not only a successful company but a successful industry.

The retail value of Scottish Farmed Salmon is today over 1 Billion and represents 40% of Scottish food exports. In 2007 Scottish Farmed Salmon was recognised by its inclusion in a list of foods in a European agreement with China that will allow a select list of British foods to be sold in China with a reciprocal agreement allowing Chinese foods to be sold in Europe. On the list is Scottish Farmed Salmon along with Roquefort Cheese and Parma Ham.

This is all thanks to those early pioneers who have made the industry so vibrant today. However, the idea is not all that new as there was a very famous fish pioneer that many city dwellers would know about. This was Lord Leverhume, one of the brothers who founded Unilever. I recently found out that his large chain of shops called Mac Fisheries was probably the first attempt to integrate the supply, processing and distribution of fresh fish.

Mac Fisheries Union Street Glasgow 1950s

It was in 1919 that he purchased the South Harris estate including the village of Obbe to turn it into a major fisheries centre. He employed over 300 local men to build piers to accommodate herring drifters, smoke houses, and a refrigeration building.

He purchased 400 shops for his Mac Fisheries chain and in 1924 the first landings arrived at Leverburgh for this is what Obbe was now called with the agreement of the population who held him in the deepest respect. Just as his enterprise got going, Lord Leverhume caught pneumonia on a trip to Africa.

Unfortunately the board of Lever Brothers had little interest in this far-flung adventure and put an end to it, laying off the entire workforce. If he had survived a little bit longer, it may not have taken so long to get started on an integrated approach to mass-market fish production. Sadly the only existing evidence of the enterprise are houses that were built for the managers.

And this is my opportunity to tell you about my own involvement in salmon fishing. The sale of wild salmon is now negligible in relation to the tonnage of farmed salmon produced. Along with grouse shooting and deerstalking it does contribute to the economy of the Highlands and Islands, particularly to isolated communities. But it is of an older, dead or dying trade that I am referring to, that of ring netting.

Netting on the Cromarty Firth at Alness

Ring netting involves hour upon hour of patient waiting staring at the water till you are goggle-eyed, looking for the surface ripple that will tell you that there are salmon running below the surface, heading for the river mouth. It may be one or it may be twenty, you will have no way of knowing but just as with 'thar she blows', a shout goes up from one of the fishers perched out on a cairn; 'fish' and all at once the person in the big wooden dinghy pulls away from the shore leaving behind a ribbon of net played out by another fisher on the stern.

As the boat speedily circles the incoming salmon, two or three others are carrying the edges of the net pushing it and any salmon in it towards the shore. Then there is the frenetic activity of grasping the salmon by the tail and killing them quickly with a smart blow delivered to the head just behind the eyes. It sounds inhumane and maybe was and I was a bit concerned about it for all the time that I actually did it.

For it wasn't to last long after I took part in it. The Alness Point 'station' was one of the last stations whose permits were all bought up by the fishing syndicates or landowners in order to protect the salmon for the well-heeled hobby fishermen from the south who pay good money for such things.

Fishing on the Cromarty Firth

The Klondykers

Klondykers in Lerwick Harbour

There was another form of fishing that went on for a number of years. While this expression is now related to the oil industry, it was at one time related to the Eastern Block factory ships that berthed mainly off Ullapool and Lerwick and hoovered up all the mackerel that the local boats could bring them. This was really in my early days in the North but I do remember them bringing a bit of colour.

Now this was in the days before the Berlin wall came down and if you are to believe Martin Cruz Smith in Polar Star, the factory ships all had commissars and spy gear. They were certainly under very strict control and were unlikely to make a break for it during their on-shore leave which was well supervised. And anyway, when you got to Lerwick where would you go? At Ullapool, which has seen a lot of development in recent years, there was a paper shop opposite the quay. Incongruently its windows were filled with radios, toasters, irons and other things you can't get behind the iron curtain. (I think there is an appropriate joke in there if you look for it). Shopping trips to Inverness were also arranged although there wasn't much to buy then. More than they could at home I suppose.

I also remember a boatload of the very fishy looking men and women waiting to be collected at Lerwick Harbour to be returned to the factory ship. Apparently there was a strict no alcohol regime on the ships. I was watching the group as the open lifeboat approached the dock. One of them was gulping from a can of Tennent's. As the boat came alongside this chap moved to the back of the queue while still drinking from his can. Eventually he had to descend the ladder which he did with great difficulty as he was carrying the can as well as a big swag bag over his shoulder. He manoeuvred down as far as he could, took a last swig of the can and left it on the quayside. His fingers disappeared over the edge, the ladder was pulled down and the boat moved slowly into the sunset to join its mother ship.

Although I never saw one of the factory ships setting sail for home in Poland or Russia with its full load of processed and frozen fish, I have seen photographs and it is an amazing sight. Apparently while Ladas were the most common cars in the Eastern block, it was very difficult get parts for them and in any case, Ladas were cheaper in Britain. It turned out that entrepreneurs from among these fisherfolk would scour Britain for old Ladas and export these back to their homeland on the decks of these factory ships. The decks would be jam packed with Ladas, some even lashed on sideways to bulkheads.

Those days are gone. The Klondykers brought a certain amount of wealth to those ports at which they were berthed, as well as to the trawlers who provided them with fish. But the Eastern block moved on, the price of mackerel fell and we saw the last of the Klondykers. They were a wonderful sight while they lasted.

If you are in Ullapool you might like to think about Lupalloo which is the family music festival. Apparently Franz Ferdinand has featured along with the Ullapool Pipe Band. And it is said that the food is very classy. Ullapool also has a popular book festival as well as a guitar festival and that is where the Junior Feis Ross is held. Very cultured indeed but very down to earth.

If you like ferries then why don't you take the bus to Ullapool and then the ferry to Stornoway. You can then do a round trip via Skye to either Kyle or Mallaig where you can then take the train back to the central belt.

In the simmer dim

Distilling

One of the other great success stories of the Highlands has been the distilling industry. While all of the distilling companies will advertise the purity of the brand, the quality of the water and the barley, the truth is that the success of the whisky industry is entirely down to the protection of the Scotch Whisky name by the Scotch Whisky Association, based in Edinburgh which since 1917 Has protected the name and waged a campaign to represent and to protect the Scotch Whisky from counterfeit and from unfair barriers to export imposed by some countries.

If it were not for the protection of the name 'Scotch', whisky brands would have long ago disappeared from Scotland along with their headquarters' offices. Because it is not well known that many of the famous Scotch Whisky Brands are actually owned by multinational companies with headquarters outside Scotland. Whisky would have been made elsewhere in the world where the air and the water is pure.

The distilling industry is a geographically widespread industry with distilleries, bottling halls and warehousing spread throughout Scotland. The logistics, overheads, cost of transport and storage are enormous, given that the minimum storage time for whisky is upwards from a minimum of 3 years. But it is also a fiercely competitive market particularly since the supermarket goliaths treat the product the same way as they do potatoes or milk, demanding lower and lower prices.

All this has led to efficiency and cost savings throughout the industry. Every company that only twenty years ago had distilleries employing tens of staff operate their stills with only a handful of operators, and in one case, Allied Distilleries Miltonduff Distillery has only a single person on site. Some still houses are monitored remotely and in some blending and bottling centres, casks are now palletised, opened and disgorged all automatically, new processes that would have the old warehousemen and stillmen turning in their graves.

On the other hand, the drives for efficiency in order to drive down costs has also meant an enormous improvement in safety in the industry. For example, mixing caustic soda together in a bucket with water for cleaning stills was once a common and dangerous task given the possibility of serious chemical burns. This has been completely dispensed with due to the introduction of clean in-place (CIP) systems in which ready mixed caustic is piped into stills and vessels. The same is the case with heavy bags of yeast which once had to be manhandled. New automated systems, along with rigorously controlled quality procedures and vocational qualifications have routinely made distilleries safe places to work.

Unfortunately that did not prevent the drowning of the distillery worker in a vat of whisky. Apparently though, he had to get out four times to go to the toilet!

At the same time the industry has acknowledged the fact that the worldwide interest in the product has meant a growth in distillery visitor centres throughout the Highlands and Islands. These are generally based on a historic and sentimentalised view of the industry and some of these visitor centres are more valuable to their owners in terms of income from visitors than the actual sales of whisky from the distillery. And why not? The employment of visitor centre staff has replaced the jobs lost through efficiency savings, albeit only for the summer months.

It is a great pity that the potential for expansion of the distilling industry in the North was not recognised long ago. While much of the raw material is grown in the north, and while the very successful malt distilling and maltings parts of the industry are based mainly in the north, the bulk of the industry is actually based outwith the Highlands and Islands, in Fife, In Edinburgh and in Glasgow. Included in this are the producers of packaging – bottles, labels, cartons, the designers and the carton companies. Whilst there are some small exceptions to this, for the main part the

Highlands and Islands has not seen the possible employment and related benefits of the growth of supporting industries.

A major exception to this is the Invergordon Grain Distillery. The Distillery was set up in 1961 and was to take its place at the heart of the Invergordon Community where it still thrives and is due for expansion under its new owner Vijay Mallya. It is recognised as one of the most efficient producers of Neutral Grain Spirit (NGS), the basic component of gin and vodka, in Europe.

Invergordon Distillery has changed hands several times over the past years and to some extent shows the dangers of globalisation and private investment.

Invergordon has been at the heart of developments in manufacturing and processing for a long time probably due to the presence of Invergordon Distillery which gives the town and area a manufacturing edge although it is certainly helped by a deep-water anchorage. It certainly gives the town its traditional smell of cooking 'mash' for the production of grain whisky. The anchorage also provides a refuge for oilrigs awaiting repair or just for storage between jobs.

Oilrig berthed under a bridge south of Invergordon

Given the enormous amount of literature available on Scotch whisky I decided that it would be unnecessary for me to say much about it in this book. Except to say that in all the years I have worked within the industry I have found everyone warm, generous and a delight to work with. It was a great pleasure knowing you.

There is one story, however, about distilling that I would like to tell you and that is about Ferintosh which could have been one of the most famous names in the world had it not been for changes in the tax system. I was very interested in the history of Ferintosh as we built a house there and lived in it for over 20 years. What I discovered was that during the eighteenth century, whisky was often called 'Ferintosh' throughout Scotland. This was because for almost 90 years, Duncan Forbes of Culloden was granted the privilege of distilling whisky free of duty. This 'Ferintosh Privilege' started in 1695 as the result of a claim by Duncan Forbes of Culloden for damages to his estates amounting to £4500 sterling. Apparently the dastardly Jacobites had ransacked his estates including his 'brewery of aqua vitae'. This privilege went on until 1784 when it was withdrawn with £21,500 paid in compensation. Not bad eh?

Because of this freebee, whisky was often called Ferintosh rather than aqua vitae which was corrupted to 'whisky'. At the demise of this cheap whisky many people were not amused and it prompted Robert Burns to write a protest song which included this:

Thee Ferintosh! O sadly lost!
Scotland lament frae coast to coast!
Now colic-grips an' barkin hoast,
May kill us a;
For Loyal Forbes charter'd boast
Is taen awa!

Original Ferintosh Old Scotch label

By the way, while the original Ferintosh Distillery (or distilleries) were based around the Ferintosh area on the Black Isle, a subsequent distillery was based in Dingwall and this area also became known as Ferintosh. The Name 'The Old Ferintosh' still belongs to Whyte and Mackay and has been produced in limited quantities.

Being Glaswegian I would I have though that Ferintosh would very quickly have been converted to: 'Gie's a wee Ferry', or 'Whit aboot a wee Tosh?' I think that would have been better than: nip, dram, nippy sweetie, hawf etc.

While we are in Ferintosh I am going to tell you a wee fact that would interest Glaswegians. Some of you might remember the pissoirs that they had in Glasgow. I do remember one in particular at Glasgow Cross. Rather than describe it I was astonished to find a perfectly preserved example behind the Ferintosh Free Church. These pissoirs were made at the Saracen ironworks in Possilpark in Glasgow. Inside, embossed on the cast iron it says: 'Gentlemen, please adjust your dress'! Cool or what?

And by the way, Edwin Morgan our own Glasgow poet knew of Ferintosh for in his poem 'Canedolia' he refers to it:

'tell us about last night
well, we had a wee Ferintosh and we lay on the quirang, it was pure strontian!....

Edwin Morgan was the first to hold the post of 'Scots Makar' created by the Scottish Executive in 2004 to honour Scottish poets. He was a worthy recipient, loved as he is by Glaswegians. You can read Canedolia and other great poems in a book dedicated to him and with comments by many contributors: *From Saturn to Glasgow.* 'Puredeadbrilliant'.

Pissoir at Ferintosh

I'm on the train!

Take a look up the rail track

I have to confess to having an interest in this as I am an anorak as far as railways goes. I don't collect numbers anymore although I have been seen to take the occasional photograph of an unusual train at the end of a platform.

Travelling to the Highlands and Islands from the South should be a pleasurable experience for both resident and tourist. I have a particular sympathy for the tourist from the Orient, particularly those four or five ladies whom I have accidentally disturbed at their ablutions over a period of years. It has always been a mystery to me why trains designers put electronic controls on toilet doors when simple systems work. This would have prevented embarrassment although I must say you get used to it and I often expect people in the loo irrespective of engaged lights being off.

Talking of the lack of foresight, I also remember the time a few years ago just after they had introduced the direct journeys from Edinburgh and Glasgow to Wick and Thurso. The same train made the journey from Wick and Thurso to Inverness and carried on to Edinburgh or Glasgow. The strange thing was that the first class compartment while in the far North was not first class but standard class. It became first class in Inverness possibly when they gave you a free newspaper. Anyway, I was getting this train from Inverness and I walked along the platform. It was mayhem. There were a number of policemen and station officials both on and off the train.

Apparently the railway people didn't know or hadn't thought about the implications of the 'T in the Park' music festival at Kinloss about to start that night. In consequence, the train was full of young folk from Caithness and Sutherland having a massive party well supplied with booze and ghetto blasters.

'I wouldn't go in there' said a policeman to me when I started to board the train.
'But I'm going into first class' I riposted.
'I particularly wouldn't go in there' he answered.

He was quite right as when I finally got on to the train, the policemen were busy removing ghetto blasters and moving young people from the first class compartment. It was a disaster of empty bottles, crisp packets and mashed sandwich. It was also a very uncomfortable journey and par for the course at that time. For years I have wondered why the Local Authorities in the North and other agencies did not work out that improving the travel to the North might actually bring people there. But I am glad to say that the takeover of Scotrail by the First Group has meant a great improvement.

As a footnote, I happened to be coming back on the train the night that T in the Park ended. It was blissfully quiet except for the snoring or groaning from the train packed with returnees from that festival who had slept unwashed in mud for the past two days. It was smelly. But they enjoyed themselves. That's the main thing!

Talking of Caithness and Sutherland, at one time British Rail maps used internally used to be upside down with Thurso at the bottom. This was because all lines to London are 'up' and all lines away from London are 'down'. Another true but useless fact.

This all reminds me of the well known buffet car attendant on the Great North Eastern train from Kings Cross to Inverness. 'Is there a Miss Murray on the Train?' 'If there is a Miss Murray would she please come to the Buffet Car as she has left something in the toilets?' How the whole train laughed.

It also reminds me of a couple of stories from the Lerwick Unemployment Benefit office on Shetland. There was a standard question on expenses claim forms that asked which was their nearest main line rail station. A wag in Lerwick said 'Bergen'. The office also applied for the use of a car to carry out home visits. As there were not enough unemployed people on Shetland, the head office suggested that they could have one if they shared it with the nearest large town – Kirkwall! Check your atlas.

Alan Gartshore also told me the story of the wee collie bitch that was being sent by a farmer in Inverness-shire to his pal, a crofter on Skye. The dug was being sent from Inverness to Kyle on the old style train with the guards van at the back. The dug had a wee tag round its neck just like a war-time refugee. Arriving at Muir-of-Ord, the guard got out to wave the train off. At this point the dug saw its opportunity to make a break for freedom and jumped onto the platform to be pursued by the guard who shouted at the porter along the platform: 'Stop that dug, He's a parcel'.

Glenfinnan viaduct Neil MacLeod

I am well used to the eccentricities inherent in travelling within and outwith the Highlands. But, with my wealth of experience I was ill-prepared for my conversation with a railway chap in Inverness when I told him that I wanted to go from Sneckie (Inverness) to Glasgow, travel to Auld Reekie from Glasgow and then come back to Sneckie from Edinburgh.

'You'll be wanting a ticket to Shotts then' he said. 'But I'm not going to Shotts'. I said'. I don't have a particular antipathy towards that famous mining town and those familiar with my previous excursions into the industrial hinterlands of Central Scotland will know that I speak highly of the Shotts Leisure Centre.

'That's all right' he said' 'Shotts is mid-way between Edinburgh and Glasgow'. 'That may well be', I opined. 'However, I am not travelling on that line' knowing well that the train for Glasgow Central to Edinburgh via Shotts takes twice the time of the Queen Street service as it takes you on a rural excursion through North Lanarkshire, lovely as it may be in parts.

I believe that he muttered something like 'Trust me, I'm a railway clerk'. So I purchased said ticket and it was with trepidation that having travelled through Glasgow and arriving at Haymarket (not via Shotts) I was accepted through the barriers without question. To preserve the issue for posterity, I have kept the ticket.

I never ever, ever travel without earplugs. So the wailing children, the MP3 player and the loud English boomer who has discovered the Highlands and is telling everyone about it don't bother me as much as they used to. However, when you have been away a few days and want to just be left alone in your seat it can get frustrating when someone wants to sit beside you. If you are gentleman like me of course, there is no point in trying to play 'the train game' by filling the other seats up with your stuff. You end up offering to move things anyway.

The railways in the Highlands are wonderful. You must take the train from Glasgow to Mallaig via Fort William. Cross to Skye and come back on the Kyle Line via Inverness. Take your time about it and you might even get on the Harry Potter Express as it has come to be known: The Fort William to Mallaig Steam Train. Simply wonderful.

I love going from Edinburgh up the east coast through Kirkcaldy and on to Perth and Inverness. I love this poem:

THE BOY IN THE TRAIN

Whit wey does the engine say toot-toot
is it feart to gang in the tunnel?
Whit wey is the furnace no pit oot
When the rain gangs doon the funnel
What'll I hae for my tea the nicht?
A herrin' or maybe a haddie?
Has Gran'ma gotten electric licht?
Is the next stop Kirkaddy?

Is yon the mune I see in the sky?
It's awfu' wee an' curly,
See! There's a coo and a cauf ootbye,
An a lassie pu'in a hurly!
He's chackit the tickets and gien them back,
Sae gie me my ain yin Daddy.
Lift doon the bag frae the luggage rack
For the next stop's Kirkaddy!

(Extract) MC Smith 1869

See eagles!

It was just the start of another ordinary day in the West. Alan and I had started out early from Dingwall to get to Barvas on Skye to carry out some training for the directors of the Community land buy-out there.

We got to Barvas and after some small diversions arrived at the community office, which, with some re-arrangement would serve as a training room. There was no one around so we hung about and had a cigarette till in about 10 minutes Kenny arrived round the corner to greet us.

The door being opened we got in and started to set up as the rest of the 6 participants arrived. 'Is such and such coming?' 'Oh I don't know, I'll give him a ring'. 'Did anyone get milk?', 'I'll just go up and get some then'. 'Has anyone done anything about lunch?' 'Oh I don't know. I'll have to see'.

The second last arrived and we got going, closing the door which opened directly on to the cobbles of the steading. As normally the last arrives just as you are starting, he did. We couldn't get the door open.

Not our place to interfere but four strong men couldn't do it. 'This way Kenny. No, No, push!' says Angus 'Let me try'. It had closed on the latch and couldn't be opened from the outside. We were in a quandary and there was consideration given to breaking open the door. 'Just pass the key through the window till we see' says I. The key was duly passed and the door duly opened. Embarrassment was glossed over and we attempted to get going.

We managed to work till lunch when an enormous plate of sausages on rolls arrived on a tray. While we were about to re-start a little girl appeared round the corned of the steading pursued by two lambs which took the opportunity of the open door to rush in and round the floor hotly pursued by the girl, in the process scattering papers, chairs and books. It took us some time to shepherd the lambs out and regain the room for our own use. While we got going again I enquired about the ownership of the lambs that turned out to be future lunch as they were heading for the shipping container slaughterhouse right up against the steading wall. We had just eaten their siblings. Very nice too!

Right oh. So we were in full flow and making progress when there was a knock at the door. Everyone looked with trepidation at the door as it was forcibly screeched open. There was Mr and Mrs tourist birdwatcher complete with binoculars. 'I'm awfully sorry', he said nervously on looking round the assembled group 'Is this where we see the sea eagles?' Just another day in the West.

Sheep

First of all, there is the great sport of sheep avoidance which can be practised with alacrity particularly in the spring and summer months.

They are generally quite stupid so the unwary traveller must at all times be on their guard particularly on the unfenced roads in the West. Apparently it is cheaper to allow sheep to be killed rather than fence thousands of miles of open moor. I don't want to open any kind of debate but is that not cruel?

Anyway there are kamikaze sheep around and these are the ones to watch for. I encountered a particularly sneaky specimen between Broadford and Portree one fine sunny morning. The sheep was grazing on some concrete on the middle of a bridge as I approached.

I saw it and prepared to slow down by putting my foot ever so slightly onto the brake. Now seeing me, the sheep darted immediately to the other side of the road where it stopped suddenly. You know it is amazing how fast sheep can move and how they can stop suddenly. I sped up thinking that it had its moment and had gone back to chewing the concrete. It got the better of me and suddenly shot back to the middle of the road whence I again slowed down preparatory to evasive action. It careered towards the side of the road and just as I was sure I could pass again it shot back straight towards the car where it broke itself as well as my number plate. As I was going so slowly I swear it was its own momentum that did it.

At least it had bounced off the road and was now lying against the wall of the bridge. I didn't bother taking its pulse. So I got on the road. 'What!' you say. 'You should tell someone!'

Like who? There is a strange rule I believe that you should inform the police if you kill a dog but not a cat, a cow but not a deer. Who knows and who cares? If we waited for every sheep to come to a full stop before getting on our way there would be no traffic movement at all. If we informed local crofters about deid sheep they would get no rest and you would get no thanks. Just take care.

And while we are on the subject of sheep, it reminds me of the story of the new primary school teacher who had come from Edinburgh to teach in a small primary school in Sutherland. She had been there a good wee while and thought that it was time to smarten up the otherwise dowdy classroom with some rural scenes. She duly sent off an order for some posters.

When they arrived, she gathered the class of small people round and spent a jolly morning agreeing where the various posters of cattle, sheep, and hens and so on would go on the walls. The posters being duly displayed she though that was a very good opportunity for a lesson in rural affairs.

'Now boys and girls' she ventured 'now that we have the nice pictures up I wonder if we can identify them all'. She pointed first of all to the poster of the sheep. 'Now then' she said, what is it that we have here?' Silence and some quizzical looks is all that she got from the small class of six whom she thought were actually very bright.

Minutes passed or seemed like it as the bairns looked at the picture. 'Come on now Alistair' she said to a young lad whose father, she knew was a crofter. The little boy looked at her pityingly and turned to his classmate. 'What do you think then Lachie? I think myself that it is a Perth type Blackface because the coat is very heavy but it could be a Galloway right enough and not very good for overwintering in these parts. It would probably do well in….. '

The posters were, over a period of time, replaced with some very nice art prints.

In the same vein, I was told about the school inspector who visited Tore Primary School on The Black Isle and after asking a class a number of very stupid questions he said to the class: 'Now boys and girls, what age do you think I am?' There was a long silence till wee John's hand shot up: 'Please sir, 38'. 'Well done boy. How did you guess that?' 'Please sir, my big brother is 19 and my mother says he's half daft……..'

'Now class, 'p' is silent as in ?'

Wee Jimmy: 'Bath' miss?

Shetland Crofthouse

The school ceilidh

What a wealth of talent there is in the Highlands and Islands, much of it promoted by the Feis movement and such organisations as the Caithness Music Festival and the Orkney Folk Festival not forgetting the Battle of the Bands and all the town and village galas and Highland Games.

Fundamental to the promotion of talent is the school ceilidh, that once-a-year bash at which the pupils of local primary schools get to show their talents. For many years there have been local battles for the retention of small primary schools while councils have been in favour of bussing and concentration of pupils in larger in larger schools.

While there are often good practical reasons for closure when there are few pupils, the local primary schools along with their associated village halls provide a focus in the community for both parents and children. We had a real battle for the retention of Kinkell Primary School at Ferintosh, which had been threatened with closure and the pupils transferred to a new primary school to be built at Culbokie, on the Black Isle.

Anyway, the school ceilidh is necessary and looked forward to and prepared for with meticulous care. The acts depend on the available talents within the school and local area at the given time. And there has to be a 'fear an tigh', the master of ceremonies, usually a local and verbose worthy who introduces acts and tells a few jokes.

The preparation involves everyone in the area. Fathers are pressed into service to move the heavy tubular and canvas chairs into rows with minimum space in between. Mothers (that's how it is) prepare egg sandwiches and big slices of clootie dumpling. Large tea urns and kettles are cleaned out. And the children rehearse in the hall and in the school.

One particular year the fear an tigh was Argo Kennedy in his Andy Stewart regalia who started off the proceedings with a glass of lubrication in his hand. One by one, the acts come on and how gallantly and stoically do we praise them for we love them dearly. Lindsay on the fiddle, Kirsty and Mary on the clarsach, poetry reading by the infants and so on with a first half finale by the policemen from Gilbert and Sullivan's Pirates of Penzance. It was very funny.

Then there is the intermission in which allocated mothers and children pass round huge aluminium kettles to fill dainty blue china cups which we have to lay on the floor in order to hold the paper plates full of a selection of egg sandwiches, sausage rolls and dumpling. It is a great balancing act and all carried out with the maximum mayhem in which juice is spilt, children climb over cups of tea and dumpling gets stood on. Some child has over egged and is sick. More tea is forced on us and the remains of the sandwich plates are handed round for those brave enough to want more.

And then the second half starts and this is where you need the excuse to pop out for a long smoke or a breath of air. This is because you know that the second half is going to be identical to the first half except in reverse order. Wills to live have been lost. The end being reached it is then absolutely necessary for parents to console one another with the gin and tonic or large dram in the nearest parent's house.

Now the ceilidh is a traditional way of enjoying yourself. Tourists coming to the Highlands and Islands may be enthralled by the accordion and kilt in the corner of the hotel lounge but most people in the North would not identify with such a spectacle. Many festivities in the Highlands happen outwith the tourist season. Get-togethers such as the regular dance at the Culbokie Hall draw together many people from the local community. These festivities are extremely well prepared for. Wines and beers are selected, as are dips, snacks and appropriate table settings, some going as far as the crystal glasses and the candelabra!

We gather at the hall and the ceilidh band commences and we all take part in dashing white sergeants, eightsome reels and the like. Even Glaswegians with big feet can learn the steps and once learned you can keep learning and improving as there are many dances and ceilidh bands are what normally play at weddings.

There is a break during which we lose a lot of money in a good cause trying to get your one pound coin closer to the wall than the others. The closest gets a proportion of the winnings. A raffle is drawn and you win a bottle of shampoo or an egg whisk.

The band re-starts and we wend our way into the wee sma' hoors * until they play 'The Last Waltz' and 'We're no awa tae bide awa' brings us together in one great circle if we can manage. Feet will never be the same again. On one occasion that was the only song I heard and I am still reminded of it.

Dances are often held on a Friday night as going on into the early hours of a Sunday is frowned upon. Anyway the local dance was on the Friday and I was looking forward to it. I had been in Orkney and was due to fly back from Kirkwall to Wick, pick up my car at Wick airport and drive straight to the dance.

So there was I sitting in the lounge waiting for my flight to be called and suddenly wondering where was my waxed jacket? Oh dear. There it was pictured by me hanging on a hook on the wall in Kirkwall College. But, my car keys were in that very jacket and my car was at Wick airport. I had enough time to call the nearest mainland Renault garage which was in Thurso. They appraised me that as my car had electronic locking it would cost an arm and a leg. They could get the car open but they would need to re-programme a key. Blah and so on. The next thing I did was to phone my dear one just as the flight was called.

It's hardly any time from Kirkwall to Wick and I arrived with trepidation to be told that she would have to get the key on the bus but it would not arrive till 11 pm. And it was raining. And what do you do on a winter's Friday evening in Wick? You have a curry with a half pint of beer and then wander between hostelries having cokes and looking like a plonker. Miserable as sin while you know that the festivities are going on in Culbokie.

So it is that the keys arrive and I eventually set out to reach Culbokie village hall just as the band is playing 'We're no awa' tae bide awa'. No wonder they shook their heads. Heads were there to be shaken.

*wee sma' hoors
I have to point out that in this context this means 'In the early hours of the morning'. I make this distinction as in Glasgow it means 'little ladies of the night'.

Stromness Orkney

Roman in the gloamin

The Roman Road, Mulchaich

Of course, Highlanders party as much as anyone, if not more. Parties, particularly at New Year when 'first footing' often last for a week. And so it was that Gordon invited us to a party at his house in Achterneed, that windswept hill station between Strathpeffer and Dingwall. Except that it has the very nice Neil Gunn Memorial*. Anyway, it was a themed party and we were to go as Romans. Now themed parties are not particularly the rage but it is a fightback against the rest of the country who have started to go to parties and rugby and football matches in Highland gear.

So a lot of preparation went into finding and fabricating togas and other Roman stuff. And it was a fine party. The Roman senate surpassed themselves as did the slaves and vestal virgins. Now, it being May, it was still winter outside and the curtains were closed. It was around 5am that we thought it was time to go. As far as I remember the children were just old enough to be left but for us (i.e. me) it was still a worry. I looked out of the curtains and got a shock. There was about 3 feet of snow on the ground. While it had been very cold when we got there, this was a surprise. Maybe it shouldn't have been a surprise as I have seen snow in June on Drumochter Summit on the A9. It had got light by then and so we found ourselves virtually snowed in but keen to get home. Luckily Centurion Ivor had a big Chelsea chariot and cleared the way downhill

The great outdoors

I hurt my back. We had moved into a little cottage kindly supplied by Doug Lawson at Lentran, his farm near Beauly. We had just moved in and I was clearing an area at the back in order to corral our two little children in order to protect the sheep and other farmyard animals. I thought to lift a log and it was only after I had started to lift it that my health and safety training came back to me spurred on by an immense shooting pain up my spine. The log had been there for years and not only was it waterlogged but virtually glued to the ground by weeds. I put my back out and was laid up for at least a week, unable to move. I was not a well man.

So there was I, a miserable city lad, sore back, sorry for myself and honest to God when my wife suggested that I go for a walk I actually said: 'I am so bored. There is nothing to do out there'. I am embarrassed now to remember it as the Highlands and Islands is one of the most exiting and adventurous places to live.

I got to love the outdoors and became a founder member of the Dingwall Ramblers (3 Paid up members). For many years myself, Henri, Douglas and occasionally supplemented by others would merrily hike through Ross-Shire's lovely hills and glens at first accompanied by Henri's faithful red setter, Yosser who sadly passed away.

From time to time I would accompany another group of serious walkers on their serious walks further up the serious hills and further up the serious glens. These were the Blistering Pacemakers whose mission was to have a jolly time but also to raise money for charity through their management of the great Cross-Ross walk from Dingwall to the West. Now in this group was a real character, Stewart Christie who had cancer for many years but this rarely stopped his hillwalking, Munro bagging and charity activities. He was both a good friend and a member of the family. It was after he passed away in 1998 that I was asked to carry out an amazing engineering feat. Along with others I was recruited to climb to the top of Ben Wyvis carrying the makings of a memorial. I carried a container of water, someone carried the small, engraved stone and others carried sand and cement.

His dad Jimmy and his wife Dorothy had been taken by helicopter to the top. Hidden in a corner on the side of the Ben is the memorial to 'A free spirit' and it was there that Dorothy scattered his ashes and we watched them drift down the mountainside on such a beautiful day. Now, also carried up the mountain was a quantity of good malt whisky and this was planted behind the stone to sustain the weary traveller. From time to time those hardy souls take to the hill and toast his memory. He is missed and well remembered.

It was a pleasure to be asked by the local Secondary School (wherein were my offspring) to supervise Duke of Edinburgh Award hikes that they now call expeditions. Myself and pal Wendy and others had a great time doing this.

Stuart and Dorothy Christie

Anyway I was supervising a group of Bronze Award pupils on their expedition over one day away above Beauly. I was keeping out of the way as I was supposed to do but I could see one of the young lads struggling at the back. His starting weight was already a disadvantage as was his constant complaining which was not endearing him to the rest of the group.

Although the group was supposedly briefed and prepared, I wondered about this and came up to them when they stopped for a break. Sneakily I lifted each of their rucksacks. They were all fine except when I got to that of the lad who had been lagging. In risk assessment for lifting and handling you first of all test the weight of an object by lifting it very gently and never assuming that it weighs what it looks. So I made the mistake of assuming the young lad's bag would be of the same kind of weight as the others. This mistake nearly cost me my arm as it was on the heavy side for a young lad.

'Do you mind if I have a look in your bag' I said to him quietly in order not to embarrass.

I got him to open the bag and the first thing that emerged from the top was a 10 pack of custard creams. Now bear in mind that this was a one-day expedition and they were expected to take their own lunches and snacks. 10 custard creams while not being particularly nutritious might have kept me going for a whole day. Anyway, out they came to be followed by:

A full family size loaf
2 family size tins of HP Baked Beans
A plastic litre bottle of milk
A small kettle
Mixed lots of tea bags etc.

I gave up counting when the entrenching tool was produced. An entrenching tool is basically a fold-up spade suggesting as it does that it is used for 'digging in' when under fire or for burying your poos. They are fairly lightweight as aluminium goes but still heavy enough to take you over the edge.

'Who helped you pack your bag?' I queried?

'My uncle' he said 'He's a survival expert' he added.

'Hmmm' I said.

'Oh dear' I thought. I could hardly interfere in and destroy what was clearly a fine mentoring relationship between uncle and nephew.

'Is your uncle good at map reading?' I asked?

'I think so' he replied.

'Well. I tell you what. I think we should bury these things, take a map reference and your uncle can come and dig it up'.

So I proceeded to dig a sizeable hole and therein interred a week's supply of groceries like the treasure of the Sierra Nevada. They are near the dam just above Beauly. The wee lad finished the walk but it had been a bad start and quite an effort for him. I was appraised later about the uncle and the fact that on his practice hike the lad had taken a two-way radio to be in touch with his mentor. It reminds me of The Survival Shop in Buchanan Street in Glasgow that closed down.

You might think that there's no excuse for this kind of preparation. But I have to tell you that on our first overnight hike in the Highlands I forgot all my Boy Scout training for there was Henry and I huddled in black plastic bin bags watching a comet shower on the night of the 1 August. It was zero degrees and my camping gas had run out. We shared cold sausage.

I am afraid this is mild compared to some of the sad stories of those coming from the south to climb some of the highest peaks in Britain. You can hear them in the bars of Fort William blithely talking of how they will 'walk in, traverse that peak, take in four Munros and a Corbett or two' and so on. The sad fact is that up to fifteen dead bodies are taken off the peaks in Lochaber alone each year.

So if you come to walk in the Scottish Mountains then do it properly. Get advice from the right people and walk within your limits. And please do not try and walk the West Highland way at one time if you have never walked long distances before. Do something shorter. Do part of it or the Fife Coastal path or St Cuthbert's Way, or one of my favourites, the Speyside Way which goes through fishing and distilling country. You can also start the journey on the lovely Speyside Railway from Aviemore to Broomhill.

But you know I remember that night of the 1st August so well as it was one of the most dramatic nights that I have ever seen. Without the light pollution the meteor shower went on for hours and was truly magnificent. People who never get into the country don't know how beautiful a night sky can be away from the city lights.

On another occasion, away above Achnasheen where the scenery is breathtaking I supervised a group of Silver Award pupils. The group split into two parties walking towards one another and due to meet in the middle. Our party got there first and settled down to wait on the others. Pretty soon we saw them coming over the top of the hill. Through my binoculars I could see them and I was a bit alarmed that having spotted us a couple of them appeared to be running. As I watched, one of them appeared to tumble and then to be surrounded by chums.

So, up the hill we marched quickly to be met by breathless youngsters. 'Kirsty's fallen over. She's broken her ankle'. We got to the hapless child and sure enough there she was in a sorry state. In these circumstances when a break is even suspected you don't remove the boot as it can act as a splint and stop the foot moving. What I did was to cut a length of camping mat to make an additional splint and support to protect her ankle. But I needed something to bind it tightly. There were about 8 youngsters gathered around and I asked:

'Has anyone got any string or something like that?' One young lad responded in the positive and with rather an awkward smile produced a largish roll of yellow insulating tape. We all looked at him in utter amazement and it appeared to take Kirsty's pain away for a few seconds. I started to ask: 'What on earth are you..... Oh never mind just now'. I never found out why. We settled down to wait for the Mountain Rescue to appear as we had sent someone to raise the alarm. That's strange as it wasn't that long ago and you might now say 'why didn't you use a mobile phone?' Well we didn't have one between us. But one of the girls did have a hair dryer!

Anyway, the emergency centre decided it would be better to send a helicopter and so it was that an Air-Sea Rescue Sea King descended on us for Kirsty to be wheeched away to Raigmore Hospital. I have always wondered what the Accident and Emergency Unit made of the nice red splint neatly bound together with yellow tape. And I don't have a photograph!

But you know that besides this, my experiences with the young people of the Highlands and Islands was absolutely wonderful for I never failed to be amazed at their politeness, their ambitions, their skills and their enterprise.

Of course, besides falling from 2500 feet there are other smaller dangers. Now I will not make a big thing of midges. There are worse things than midges in the wild - like Glaswegians on jet skis. It is said that if they solve the midge problem then the tourist trade would take off and we would get lots more people up North. Hmmmm.

Anyway - one story. I am not greatly troubled by midges in the normal course of events so I didn't take many precautions the day I went to Mar Lodge, looked after by the National Trust for Scotland. I was assessing people on environmental qualifications linked to the restoration of footpaths. These people are wonderful. They can work in the most appalling weathers to keep access to the wilds open. Anyway, I realised my mistake when we had walked on a beautiful day to the site.

The gang was cutting a new road through the heather and I had not experienced midges that far east. The very worst time for midges is when new ground is being opened. Anyway, each of the gang was wearing a midge hat and they were fully covered up irrespective of the blazing sun. And there was I standing back frantically puffing turbines to create smoke in order to keep the beasts at bay. It was the team leader who let me into the secret. Besides being covered by midge hats which have a net like a beekeepers covering the head, they all wear Avon 'Skin So Soft' skin cream which works best of all potions and ointments. Now in discussion with the said leader I noticed that he had an antipodean accent and on my enquiring he informed me that he was one of our colonial cousins from Australia in fact. Not only that but he was from Alice Springs.

Now in my deep wonder and amazement I enquired as to why he from a town like Alice was standing down a midge infested hole in Braemar. He smiled cheerily and said: 'Just come to see the world matey!' Nice to meet you guys. You do a wonderful job.

Oh, in those wild spots there is another little nuisance. That is the sheep tic. If you had the courage to watch a sheep tic at work it is a little beastie that latches on to your skin and then sucks your blood at the same time growing into a little blackish red balloon.

This might go on unnoticed till you become aware of it or someone else does like my daughter Catriona.

'Oooh what's that on your ear - gross!'

So you knock it off and you may be left with a little lump and maybe a little feint because of a lack of blood. Well?

Anyway you go to Doctor Munro's when you have enough ailments gathered together to justify his time. And he says 'Oh that's allowed'.

I went to him with another ailment - I can't remember what and he says: 'That's one of these things we get from time to time'.

It's a super surgery in Beauly by the way but you feel if you go along with your leg falling off they might say: 'Don't worry, that's one of these broken legs we get from time to time!'

Motorway service station near Beauly

Getting along swimmingly in Moray

Being of a fairly energetic disposition my other favourite occupation while travelling about the Highlands and Islands was to visit the many excellent swimming pools provided for and sometimes by the community. I have been in most from Stirling to Lerwick, taking in Thurso, Kirkwall and Stromness.

Strange as it may seem, the Highlands and Islands pools are not only for swimming. In Shetland they are used for sea survival courses. Kayaking takes place. Of course many of them double up as school pools but are open to the public. In this way they get a tremendous amount of use. This is the case with Speyside High School in Aberlour which I visited once around lunchtime during my distillery visits.

Now readers of my previous works might know that I have always had a love of swimming. In one's halcyon days in Glasgow, one would simply strip in the cubicle at the side of the pool, don the trunks and immediately plunge into the pool.

However, in these days of hygienic rectitude (can I say that?), one is required to shower before plunging in. Well I had not been long in the Highlands to experience this newish measure. God knows the number of small children that must have perished before it.

Anyway, I emerged from the dressing room just as a class of youths were so doing. I was eyed suspiciously indeed by a quite domineering elderly swimming mistress tagging along behind them.

I have to admit to a certain amount of embarrassment when she pointed to the sign which said: 'PLEASE SHOWER BEFORE ENTERING THE POOL' I turned round and went into the shower as I could hardly tell a lie as I wasn't wet.

On emerging back again through the doorway I looked at her standing on the edge of the pool still supervising my entrance with suspicion. At this point I saw another sign on the doorway. It said: **'HAVE YOU USED THE TOILET BEFORE SWIMMING?'** I looked at her with a 'Don't even go there' look and plunged into the pool.

Hopeman

Once a year we converged on Hopeman. Friends, mostly from Dingwall and environs set up a 'tinkie's camp' at Hopeman Caravan Park. There we were met by other friends who would stay for a few hours or a few days in their own or borrowed caravans or tents. As it was the same time each summer we got to know other regulars like the engineer from the P&O who got a prime position on top of the old platform. (The caravan site was actually in the old Hopeman Station, now grassed over). There were friends and their children from all parts and we were joined by day visitors making it a regular and fun event.

What made it even more fun was that it was generally over the period of Hopeman Gala week so that the Dingwall crowd could beat the locals in the Hall Quiz, 5-a-sides or in the tug of war. This was easily done, particularly the hall quiz at which the locals clearly got fed up at being bested by Highlanders. To overcome this they started to put in obscure facts that only locals were likely to know. Cheats!

Anyway it was all good fun. We won the raft race by Duncan pushing from underneath. There are photos to prove it! We dressed in nighties for the Tug-o-war. We always got beaten to the wellie hurling competition as we couldn't compete with those fisherfolk.

Anyway, having established ourselves for several weeks, the working elements of the party such as Dave and I had to go and earn the crust. In my case, this was by going off to work in various places throughout Scotland. During one absence, our dear son Christopher whom you have already met in Glasgow Smells through his exuberance in pushing buttons in the Perth Tourist Tank, 'found' a lobster pot 'just lying' in the grass along from the site. With his positive nature he opined that said lobster pot was a good opportunity for both fun and scientific experiment. So he and new-found friends deposited said pot in ten feet of water in a rocky inlet.

I had been in Glasgow and returned to Elgin via the train and taxi and wearing my suit. I had just arrived into the site when I was rushed by a large group of children of all ages headed up by my daughter Catriona. 'Dad, dad, dad, guess what? We have caught a lobster 'come and see' she gasped, barely able to contain herself. So there was I hauled off to this pool just like the Piper of Hamelin, but the other way around. I was hustled to the front to look down into the swirling watery pool where I could vaguely see the said lobster pot containing what was clearly a crustacean. The water looked awfully cold.

I made all the positive noises possible such as 'waow' and 'well done', but actually wanting to head for a fag and a beer. I am afraid I was ill prepared for the next statement which I can say is etched in my memory and family history to this day but has probably been the making of me.

'How are we going to get it out?' Said one young voice coming somewhere from the group of about 15 smalls gathered around the deep dark watery hole.

'Don't worry' said Catriona, 'my Dad will get it, won't you dad?' I was not quick enough to feign a broken leg or suddenly collapse in an imaginary feint. I could only stare bleakly down at the beast. (The lobster, not my daughter).

Now I love diving and snorkelling and would do it for hours in Greece and Spain. However, early in the season in Northern Scotland the sea is hardly Mediterranean and in fact is cold enough to entertain the hardiest of penguins. As they say in Glasgow:

'Many are called and a few are frozen'

So off I went, silently mulling my stupidity in the way that I do. The way that looks as if I am seriously planning how to go about a major fishing expedition, when in reality I am stricken with shock and alarm.

There is no escape when you are surrounded by a large group of expectant children and one in particular who has the highest and unwarranted expectations of your heroism. So I returned to the scene egged on by the eagerly awaiting crowd of children. I hoisted myself gingerly into the water wearing my trunks, flippers, goggles and snorkel. I knew that if I was going to do this then I had to do it in one dive. I am appraised that the survival rates for exposure in some parts of the North Sea are given in minutes so I had to get a move on. I took a deep breath as I lowered myself into the watery abyss, the freezing pool closing around me. By God it was cold but I got right to it. Actually, it reminded me of that little poem I read many years ago and has been essential in forming my philosophy:

A young man was told it couldn't be done
but he with a grin replied:
'I've never been one to say it couldn't be done,
leastways not till I've tried'
So he buckled right in with the trace of a grin,
by golly he got right to it.
He tackled the thing that couldn't be done……..
And he couldn't do it!

Anyway. I got the beast in its lobster pot and as it emerged in my hand from the murky depths it was pulled rudely from me by Indian braves who swiftly bore it away in triumph back to where the wagons were circled. I was left to pull myself out over the rocks, scraping my arms in the process and then shivering back on my own to have a hot shower in the washroom.

However, it was a huge lobster and was greeted with admiration and whoops of joy. But my wife, ever practical, was the first one to give us pause.

'How are we going to cook it?' she said. Right enough. Well the solution turned out to be really quite practical as well as entertaining. Dave and Ozzy were dispatched to the Station Hotel to borrow a large lobster-sized cooking pan. It was only after about an hour that fears grew for their safety in uptown Hopeman. I was sent to find them and lo and behold were they not finely ensconced with foaming pints playing pool with the locals, who turned out to be better at pool than at quizzes and tug o'war. So in order to maintain the spirit of the occasion it was required that I join them and it was later that the last man standing, Andrew, a visitor with a day ticket, was sent to find us. It was explained to him that the delay was in finding the pot, considering the way in which the beast was to be dispatched and of course wondering would it not be better to barbecue it anyway as both Ozzy and Dave were dab hands at barbecuing. Anyway, Andrew agreed with all and it was no trouble at all to assist his decision to stay for a last pint.

Thereupon we returned to our worried wives and children huddled in the centre of the wagon train. We entered the camp in high spirits, me carrying the pot over my head just like Ned Kelly (the Glasgow nyaff) only to be biffed and clanged by a platoon of terrorist children. 'Save yourselves' said I to the other man, 'Just leave me a gun and a few bullets'.

Anyway, was it not too late then for a supper of lobster thermidor? So it was decided that the next day being Sunday, There would be a full scale Sunday lunch, with lobster on the menu. This, I can tell you right here and now, did not please me at all. I, even though I had taken a major part in both obtaining the lobster (solo) and the pot (as a team member), cannot eat seafood and what was proposed to me was anathema as well as ge'ing me the boak: Lobster in Garlic.

Now, if any of you have been in the presence of a great chef like Teddy (of the Mart fame as previously described) while cooking crustaceans, you will well know that the living beast is dropped into boiling water in the pot at which time the air is rent with a sort of scream. It is said, probably to placate the squeamish, that this is air escaping for the beast, which doesn't actually wash with me.

So I cannot further describe the proceedings of the cooking session at which I believe a score of young people watched on avidly while this poor beast was dropped into the boiling cauldron. Being of a bloodlusty Highland disposition this was accompanied by a '10-9-8-7-6-5-4-3-2-1' and a great 'oooh' 'sordid! etc. while little ones scurried back to the safety of their caravans. This of course greatly reduced the demand on lobster leaving more to the gourmets.

And the reason I could not describe this in true detail is because the hunter-gatherers of the clan, having hunted and gathered took themselves back to the pub to recall their night of glory.

So there was then a grand buffet of lobster in garlic mayonnaise, green salad and crusty bread. There turned out to be a meaty alternative and as long as I kept out of the lobster reeking caravan I was quite safe. Wagons ho!

Catriona at Lentran, by Beauly

My dad will get it! Won't you dad?

How could I resist?

Stromness, Orkney

Also by
Michael Meighan

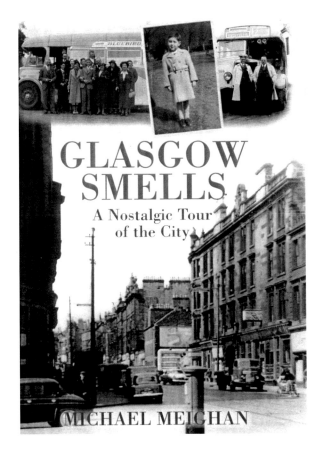

The History Press

ISBN 9780 7524 4486 4